BARLOWE'S GUIDE TO EXTRA-TERRESTRIALS

WAYNE DOUGLAS BARLOWE AND IAN SUMMERS

WORKMAN PUBLISHING • NEW YORK

Dedication

To my parents, Sy and Dorothea,
and to my sister, Amy,
whom I could not love more.
Wayne Douglas Barlowe

To my brother, Henry,
and my Uncle Ben Summers, for their love.
Ian Summers

Acknowledgments

Special thanks to Sally Bass for her support, hand-holding, and design. We thank Beth Meacham for her impeccable research and writing, and Louise Gikow, who made sense out of our scribbles. Thanks to our publisher, Peter Workman, and our editor, Sally Kovalchick, for their guidance. And most of all, we wish to thank the science fiction authors who tantalized our imaginations by creating these extraterrestrials with integrity.

Library of Congress Cataloging in Publication Data

Barlowe, Wayne Douglas.
Barlowe's Guide to Extraterrestrials.

1. Barlowe, Wayne Douglas. 2. Science fiction — Illustrations. 3. Life on other planets. I. Summers, Ian, joint author. II. Title. III. Title: Guide to extraterrestrials.
NC975.5.B36A4 1979   809.3'876   79-64782
ISBN 0-89480-113-9
ISBN 0-89480-114-7 pbk.

Workman Publishing Company, Inc.
1 West 39 Street
New York, New York 10018

Manufactured in the United States of America

First printing October 1979

10 9 8 7 6 5 4 3

BARLOWE'S GUIDE TO EXTRATERRESTRIALS

Preface

When I met Wayne Barlowe for the first time, he was eighteen. I was the executive art director at Ballantine, and Wayne had brought his portfolio over on the recommendation of a colleague of mine at Random House. I had been teaching a course in science fiction and fantasy illustration at Parson's, and my eighteen-year-old students were far from ready for the ambitious career that Wayne was mapping out for himself. A science fiction illustrator has to understand biology, ecology, climatology, anatomy, and zoology. It can take years for even a talented artist to develop the expertise needed to create believable extraterrestrial life forms. Before I saw Wayne's work, I was skeptical of his ability to do this. He was so young!

I was wrong. While Wayne looked young, his work had the maturity of a well-seasoned illustrator. We discussed every aspect of science fiction art that afternoon. Wayne was articulate and intelligent; he was also a reader, a rarity among artists. He was familiar with the works of the best science fiction illustrators, and perceptively discussed their strengths and weaknesses. But most important, he obviously was going to be a brilliant sf artist. When we said good-bye to each other, we both knew that we would eventually work together.

That day came two years later. By that time, Wayne was an established cover artist at many of the major paperback houses. He had developed an uncanny ability to grasp a crucial portion of a narrative and render it stunningly in visual terms. At twenty, he had a technical and conceptual excellence that is rarely matched. I selected some of his original works for *Tomorrow and Beyond*, an anthology of 300 paintings by 67 different artists.

And we also began talking about a collection of paintings based on science fiction literature that Wayne had been planning. From that encounter grew this book. It is a tribute to Wayne's versatility that he has been able to transcend his lush, elaborate story-telling style to create these fabulously detailed, meticulously rendered extraterrestrials. His work here establishes beyond any tomorrow that Wayne Barlowe is one of the most talented science fiction artists in the field today.

Ian Summers

CONTENTS

Foreword

Over the years, we have had lots of heated discussions about science fiction illustration. Who does what best? We've argued the merits of N.C. Wyeth, Howard Pyle, Walt Disney. Our roots are planted in the Brandywine School, in comics, and the early sf films of the fifties. Most of the popular sf illustrators trace their backgrounds to similar sources.

It is overwhelming to realize that Wayne Barlowe wasn't even around in the fifties. He wasn't born until 1958. He never read comic books, and doesn't even like fifties sf films. What makes Wayne's paintings even more remarkable is that they represent the only collection of alien illustrations that are scientifically conceived. Wayne is the son of Sy and Dorothea Barlowe, whose illustrations of animals have appeared in most of the major natural history publications. They gave Wayne an early appreciation and respect for biology and zoology, which he coupled with his own love for science fiction to produce imaginary creatures grounded in physiological realities. The elder Barlowes paint animals the way they are; Wayne paints animals that never were and makes us believe that they could be.

We met Wayne in Ian Summers' studio while we were all busy preparing and editing the manuscript for *Urshurak,* our new novel. Ian had told us about Wayne's field guide and we were anxious to see his illustrations for it. The concepts and rendering techniques knocked our socks off. Wayne and Ian had tackled a monumental subject, and succeeded at producing the finest paintings of extraterrestrials we had ever seen.

We believe that sf fans, authors, illustrators, scientists, and the general public will appreciate this book for years to come — as a beautiful art book and as a fascinating encyclopedia of alien life forms. It is the illustrated field guide to extraterrestrials that every lover of science fiction must have, and we plan to carry it with us on our next excursion to outer space.

The Brothers Hildebrandt

Introduction

When Ian and I decided to collaborate on a guidebook to extraterrestrials, we began a monumental project. I had been thinking about doing a book of paintings of science fiction characters for a while, and after Ian saw some thumbnail sketches I had done based on characters from Larry Niven's *Ringworld*, his enthusiasm for the idea fueled my own. But this book is far from the product of two overactive imaginations. We both have enormous respect and love for science fiction, and our purpose was to remain totally true to this wonderful body of literature. To do so, we got involved in technical complexities that most readers never think about.

When we began to choose the subjects for my paintings, we agreed upon a long list of stringent criteria. The extraterrestrials had to be logically and scientifically conceived. We would do as little editorializing as possible; the aliens would be painted objectively. We wanted entities that challenged the imagination, that had been created out of all of mankind's knowledge and then some. But extrapolations into the future had to be scientifically believable. The creatures could appear outlandish—the work could be stimulating and exciting—but it had to adhere to an inner emotional and biological truth.

Whenever possible, we chose aliens that had rarely if ever been visualized before. We also were careful to avoid stories that, although still great fiction, were now known to be inaccurate. A good deal of early sf dealt with aliens from our own solar system; present space probes and sophisticated new theories have rendered these creatures obsolete. We decided to only include aliens from other star systems.

We wanted a mixture of life-forms—humanoids, insectoids, reptilians. And we wanted a variety of character traits—evil, compassion, hostility, congeniality, savagery. But above all, we wanted aliens that had been conceived with dignity and integrity. We were looking for a sense that these extraterrestrials could actually exist, that they might have really evolved—and that they might be on their way to Earth at this very moment.

We soon realized that the selection process was not going to be easy. There are hundreds of aliens in science fiction, but few met our criteria. Some were poorly conceived or only partially described. Too many science fiction writers—even the best of them—take the easy way out when creating an extraterrestrial. Magic can put a cat's head on a human torso; but only science will describe why it's there, how it evolved, where it lived, what it ate and drank. We had to know all of this, and more.

We spent countless afternoons, sketchbook in hand and research materials on tap, figuring out the evolution of each alien, based on its biological type and ecological conditions. If an entity had wings, we needed to know how they functioned. Could they carry its weight on a home world with a certain gravitational pull? We discovered what our extraterrestrials ate, how—and if—they breathed, how they reproduced. If they were space travelers, how would they adapt to changing environments? If they had never left their planet of origin, how had they survived?

As time went by, we became so close to our extraterrestrials that they began to really exist for us. Sometimes, in the middle of a painting, I would look up, expecting to see a relaxed Overlord posing for his portrait in the chair opposite mine. Ian began to have strange dreams about romantic liaisons with attractive Polarian females. And the paintings kept rolling off my drawing board. I learned to reproduce every imaginable texture—wax, scales, fur, membranes, gelatinous materials. I spent hours with my acrylics, meticulously painting each and every hair of Sulidor fur with a 000° sable brush. When the work was finally completed, I felt like a member of a vast new intergalactic family.

That family is lovingly presented here for the first time. I spent many long, happy hours traveling through other worlds in their company, and I hope you find them as interesting and congenial as I did.

Wayne Douglas Barlowe

BARLOWE'S GUIDE TO EXTRATERRESTRIALS

ABYORMENITE

Source:
Cycle of Fire
Hal Clement

This top view of an Abyormenite reveals its closed mouth.

A magnification of an Abyormenite tentacle shows an extended bundle of manipulatory tendrils.

Physical Characteristics:

The Abyormenite is a genderless entity about 1.2 meters tall, with a bulbous body supported on six muscular tentacles. On the upper curve of the body is an orifice, possibly used for breathing and talking. The deep red of the Abyormenite's upper torso continues in a stripe down each tentacle, while the tentacles and the underside of the body are black. Small, delicate manipulating tendrils can be extended from the tip of each tentacle, and are retracted when the being is walking. The Abyormenites "see" by means of high frequency sound reflections.

The Abyormenites require very high temperatures to live, and an atmosphere high in nitrogen oxides with nitric acid for moisture.

Habitat:

The Planet Abyormen orbits in a complex two sun system, in a path that brings it very near the hot blue dwarf of the pair every sixty-five years. Only while the planet is near the blue dwarf are the atmosphere and temperature suitable for the Abyormenites.

Reproduction:

All Abyormenites live through the entire sixty-five-year period when their planet is suitable for them. As Abyormen retreats from the dwarf and enters the sixty-five-year period of coldness and an oxygen-based atmosphere, the Abyormenites all die, leaving reproductive spores in the bodies of the intelligent beings who dominate the planet in the cold years. When the hot years return, the cold life in turn dies, depositing its reproductive spores in the bodies of the new generation of Abyormenites. Since each race requires the death of the other to reproduce, they have arranged to peacefully alternate inhabiting the planet, neither race destroying the artifacts of the other.

Abyormenite

ATHSHEAN

Source:
The Word for World Is Forest
Ursula K. Le Guin

A gradual silvering of its mane indicates the aging process of an Athshean.

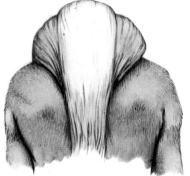

Physical Characteristics:

Athsheans are fur-covered, forest-dwelling entities. The color of an Athshean ranges from deep green to brown to black, with pale green skin on face, palms, and soles of the feet. Athshean adults stand an average of one meter tall. They have dark, glistening eyes and high, soft voices, and move with a loose, graceful gait.

It is believed that the Athsheans evolved from tree-dwelling primates after a more human species on Athshe became extinct.

Habitat:

The surface of the planet Athshe is largely covered with water, with scattered archipelagos to the south and five large land masses covering a 2,500 meter arc in the northern hemisphere. The Five Lands are densely forested with climax tree growth.

Culture:

Athsheans live in small clan groupings in communal *warrens* dug in under the roots of large trees. The omnivorous Athsheans hunt small game with spears and bows. They have both a complex spoken and written language and a highly-evolved sign language based on touch. Normally pacific, Athsheans substitute competitive singing for physical combat.

Central to Athshean culture is the concept of controlled "dreaming." Athsheans perceive two equally valid realities: "world-time" and "dream-time." Trained "Dreamers," male elders of the clan, meet for prolonged dreaming in a Men's Lodge. Their visions are interpreted by the female elders of the clan, who are the rulers and administrators of the loosely-knit Athshean government.

Athshean

BLACK CLOUD

Source:
The Black Cloud
Fred Hoyle

The brain is made up of many small units, each consisting of a piece of rock that has layers of molecular chains carefully arranged on it. These units function much like a memory bank in a computer. The units are surrounded and interconnected by the electromagnetic flow of the cloud and circulating gasses that provide energy for the units and remove wastes.

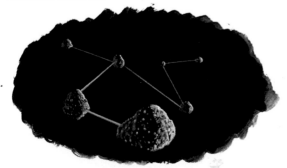

Physical Characteristics:

The Black Cloud is a vast, intelligent cloud of interstellar hydrogen, about 150 million kilometers in diameter. At its center is a complex neurological system made up of massive molecular chains.

Black Clouds travel through space, making occasional stops in the vicinity of a star in order to use the energy to produce food and the chemicals that make up their bodies. When a Cloud reaches the vicinity of a sun, it assumes a disklike shape that enables it to absorb energy more efficiently. By condensing hydrogen in a small area of the cloud, producing a fusion reaction, the Cloud creates an explosive jet of gasses that acts like a rocket, enabling the Cloud to accelerate, decelerate, or change direction.

Reproduction:

When a Black Cloud discovers a nebula of dense hydrogen gas that does not have an intelligence, it may stop and reproduce. The Cloud begins to grow a few simple brain units inside itself, organizing the magnetic flows and energy storage systems necessary to support intelligence. It then plants these units in the cloud, along with extra food energy and molecule chains, which become the nucleus of a new individual. Gradually, over millions of years, the young Cloud builds up more brain units and more complex systems until it can begin its own wanderings through the universe.

Culture:

There are thousands, possibly millions, of intelligent hydrogen clouds wandering through our universe. Their very nature makes it impossible for a Cloud to remain in one area for very long; if a Cloud stays in interstellar space, it will soon run out of energy, and die. If it tries to remain in the area of a sun, vast gravitational forces will cause the Cloud to begin to condense into a solid body.

Although individuals live solitary existences, they do communicate with each other by means of radio transmissions on the one centimeter bandwidth. These long-distance conversations are generally about mathematics, philosophy, and the nature of the Universe.

Black Cloud

CHULPEX _____

Source:
Masters of the Maze
Avram Davidson

Physical Characteristics:

Chulpex are erect, six-limbed entities that have evolved from insectoid ancestors. The uppermost pair of arms are far stronger and more developed than the lower. Their five-fingered hands have nails that are thick and yellow. Chulpex bodies are translucent, with visible internal organs. Their hair is stringy and colorless, and their pale eyes have no lashes.

Chulpex growth is stimulated by eating proteins. The worker and breeder classes are limited in height by diet to about 180 centimeters. The dominant male, or *Sire*, consumes most of the available protein, growing to an enormous size. As the Chulpex grows larger, it repeatedly sheds and regrows the vestigial carapace on its back.

Habitat:

The Chulpex inhabit a labyrinth of tunnels and caverns deep in the interior of their home world, orbiting the sun Sarnis.

Reproduction:

Chulpex are egg-laying beings, each female producing large clusters of eggs that are incubated in cool, dark caverns until they hatch. The tiny hatchlings are fed on the bodies of dead adults until they grow to the proper size for their destined roles in life. The Sire of the swarm is the only male who is permitted to fertilize the breeder females, making all members of the swarm his children.

A rear view of a Chulpex, showing its carapace. The carapace is shed and then regenerates as the entity grows in size.

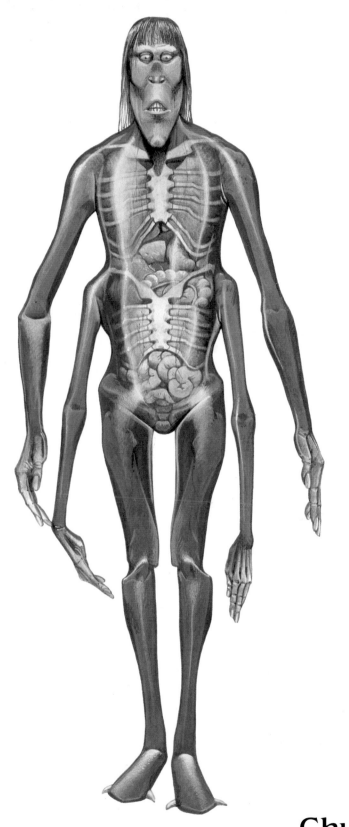

Chulpex

CINRUSS

Sources:
Hospital Station
Star Surgeon
James White

Physical Characteristics:

The Cinruss is a 1.5-meter-tall insectoid entity with small but functional wings. Sucker-tipped feet on its six legs enable the Cinruss to walk on walls and ceilings. It also has four manipulatory appendages that are capable of extremely delicate work. The Cinruss's beaklike mouth is centered between its faceted eyes.

While the Cinruss is not telepathic, it is strongly empathic, sensitive to the emotions and sensations of all other beings.

The Cinrusskin are oxygen breathers, and have a versatile metabolism. They are able to obtain nourishment from most foods grown in a carbon-oxygen environment.

Culture:

The Cinrusskin evolved in a low gravity environment, and must wear an antigravity device whenever they venture into a higher gravity situation. Their fragility has given them a sharp sense of danger, and caused them to develop rapid reflexes. Because of their empathy and their aptitude for delicate manipulation, they often become physicians or surgeons, specializing in interspecies medicine.

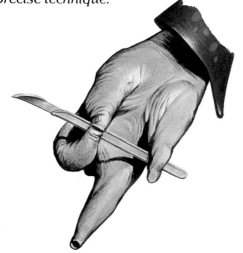

A sucker-fingered Cinruss hand holding a surgical scalpel. Cinruss surgeons are known galaxy-wide for their delicate, precise technique.

Cinruss

CRYER

Source:
Conscience Interplanetary
Joseph Green

Physical Characteristics:

The Cryer is an independently functioning unit of a planetwide silicon-based plant intelligence inhabiting the planet Crystal. The Cryer resembles a two-meter-high bush with a crystal and metal trunk and branches, and small, sharp glass leaves. The trunk contains silicon memory units, powered by a low-voltage solar storage battery and connected by fine silver wires.

Located about 180 centimeters up the Cryer's trunk is an organic air-vibration speaker membrane created for it by the planetwide entity to enable it to speak with human beings. It is a broad, saucer-shaped leaf held in place by stretched wires to provide a vibrating diaphragm. A magnetic field generated in silver wire coils hanging on either side of the speaker causes it to vibrate to produce sound.

The planetwide intelligence, which calls itself Unity, consists of thousands of smaller units like the Cryer, connected by an underground nervous system of fine silver wire. Each unit has a specialized function, some storing electricity generated by sunlight, some extracting silver for constructing the nervous system, some providing memory storage, and some acting as sensor units. Unity is able to perceive temperature, motion, position, electrical potential, and vibrations through its member units.

Unity and its organs and subintelligences become unconscious while absorbing sunlight, sleeping during the day and working at night.

Habitat:

The planet Crystal's atmosphere is 18 percent oxygen, with nitrogen and hydrogen making up the balance. Life on the planet is based on silicon, with a high proportion of metallic elements.

The air vibration speaker membrane of the Cryer was created by Unity to facilitate communication with Terrans.

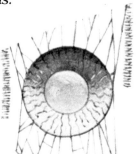

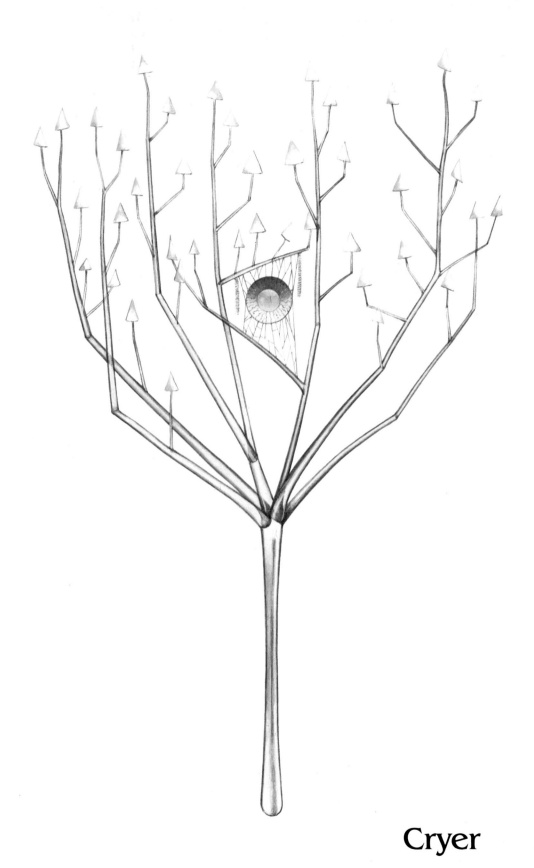

Cryer

CYGNAN

Source:
The Jupiter Theft
Donald Moffitt

Physical Characteristics:
The Cygnan is about 1.5 meters tall, with six limbs that can be used as either arms or legs and a long, three-petaled tail that folds to conceal the sexual organs. The slender, tubular body is built on a cartilaginous skeleton, with the brain located between the upper pair of limbs at the top of the spinal cord. The three eyes are placed on stalks in an equilateral triangle around a broad, flexible mouth. The Cygnan has a harsh, rasping plate in the mouth, and a spiked, tubular tongue.

The Cygnan nervous system is extremely well integrated, with much faster synaptic reflexes than those of a human being.

Habitat:
The Cygnans evolved on the satellite of a gas giant planet orbiting a double star. In time, one of the members of the stellar pair collapsed into a black hole. The Cygnans had sufficient warning of this impending catastrophe to be able to construct huge spaceships for escape. The entire Cygnan race now lives in five fifty-five-kilometer-long spaceships. The interior of each ship, aside from the purely technical sections, is a huge, open, artificial forest; here the Cygnans live alongside the small arboreal animals they catch for food.

Culture:
The Cygnans' long isolation inside their ships and their lack of interest in anything that does not directly affect their own survival has developed in them an absolute disregard for other intelligent species. They have a very sophisticated technology and are completely at home in space.

Their speech is musical, consisting of chords produced by their multiple larynxes, and depends on absolute pitch. Their language is incredibly rich and varied; they have more than a million phonemes available to them, and each word is made up of several phonemes.

Reproduction:
The Cygnan pictured is a female. From the sketchy information available to Terran scientists, it is hypothesized that the small entity located on the torso is the Cygnan male, an apparently parasitic organism. The method of reproduction is presently unknown, but the male remains with the Cygnan female for life.

The tubular, spiked tongue helps the Cygnan to produce its beautiful musical language.

An engorged parasitic male, showing his extended feeding tube.

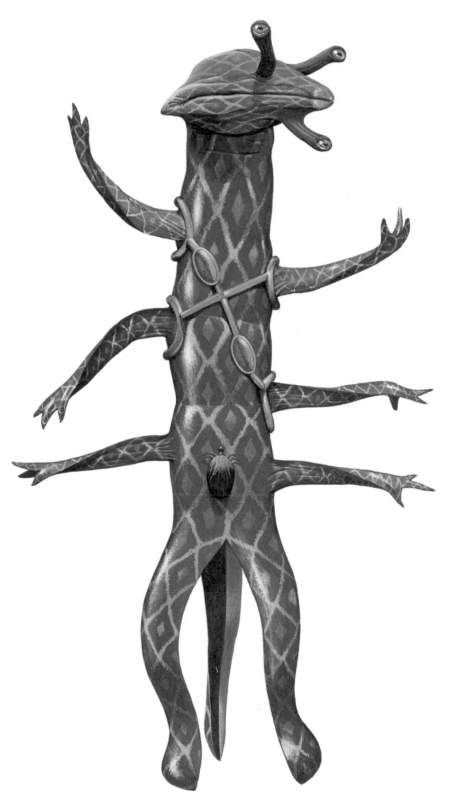

Cygnan

CYGNOSTIK

Source:
A Little Knowledge
Michael Bishop

Physical Characteristics:

The Cygnostik has two slim arms and legs, a cylindrical body covered with a ragged, shroudlike growth, and a narrow head bearing a halolike crest of bone or cartilege. The multijointed legs may be either extended or retracted. In the retracted position, the Cygnostik stands approximately two meters; extended, he is 2.5 meters tall. While the body and head seem to be entirely organic in nature, the limbs have the appearance of sophisticated prosthetic devices.

Each eye contains two horizontal pupils connected by a thin rod, forming an hourglass shape. One pupil faces forward; the other is located on the outer edge of the eyepatch.

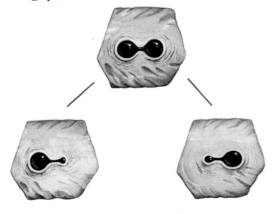

The face and body are of a mahogany color that grades into the metallic gray tone of the limbs. Parallel nose ridges end in a single nostril above jointed mouth parts. The eyes look like patches of wet canvas, with four pupils whose color varies widely in individuals from blue to orange to amber.

The two front-facing pupils function very much like human stereoscopic vision, with a selective sensitivity to the spectrum of the Cygnostikoi planet's primary star, 61 Cygni A. The peripheral pupils operate independently of one another, and have their retinas attuned to the radiation distribution patterns of the smaller sun in the binary, 61 Cygni B.

During religious observances, the frontal pupils narrow to tiny pinpoints, almost disappearing. The side-facing pupils widen and glow brightly. It is conjectured that the peripheral pupils allow the Cygnostik to apprehend a spiritual realm that humans seldom perceive. The Cygnostikoi can look into both the spiritual and physical worlds at the same time, and often do.

While on Earth, the Cygnostik's diet consists of simple fruits and meats; it has a marked fondness for cat. Its metabolism is so efficient that it leaves almost no waste products, except a clear, vinegary-smelling viscous liquid excreted through microlasered pores to lubricate the limbs.

Cygnostikoi speak a musical language, which humans are able to reproduce only on a synthesizer.

Habitat:

Little is known of the Cygnostikoi planet of origin, which orbits the binary star 61 Cygni. From their controlled living environments on Earth, it is suspected that their home world is one of dim and continuous twilight, with temperatures seldom exceeding 0° C.

Culture:

Cygnostikoi on Earth live in a seven member conjugal bond. There are no reports on their sexual behavior.

A striking feature of Cygnostikoi culture are its worship practices. One member of a Cygnostikoi bond becomes the object of worship each day. Worshipers tear off the ragged strips of hanging flesh from their own forelimbs and feed it to the object of worship. This member collapses at the end of the worship service in a cataleptic trance resembling death, not reviving until the following day.

Cygnostik

CZILL

Source:
Midnight at the Well of Souls
Jack L. Chalker

A cross-section of a Czill foot, showing one of the two Czill brains and the tiny drinking tendrils.

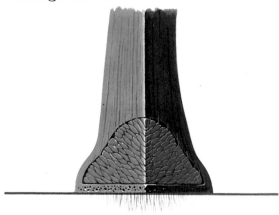

A view of the extended drinking tendrils, ready to absorb the water necessary for a Czill's survival.

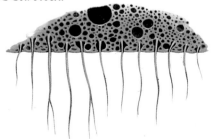

Physical Characteristics:

The Czill is a mobile plant intelligence more than two meters tall. The smooth green skin is covered with transpiration pores that absorb carbon dioxide from the air. The Czill's long, flexible arms and legs have no joints, but are able to bend at any point. A large, stiff leaf used for photosynthetic food production is attached to the entity's head by a short stalk. The mouth is a broad slash used only for speaking, and the single rigid nose hole is used for detecting odors. The unlidded eyes are always open.

The Czill has two brains, located in the broad, round feet. Each brain controls one side of the body, but they are connected to correlate information and for memory storage.

Every few days a Czill must absorb water. The organism enters a shallow body of water and extends tiny rootlets from the bottom of each foot. Water passes through the rootlets and into the body.

The Czill takes root wherever it is standing at nightfall and loses conciousness, remaining in a pleasant, dreamlike state until the first rays of sunlight fall upon the leaf.

Reproduction:

An individual Czill commonly lives about 250 years, and in that time will reproduce itself about four times. Over a period of about ten days, the Czill fissions into two identical beings, each one with all the memories and personality of the original. Czill custom decrees that the twins immediately separate, so that they can have different experiences and so develop into different entities.

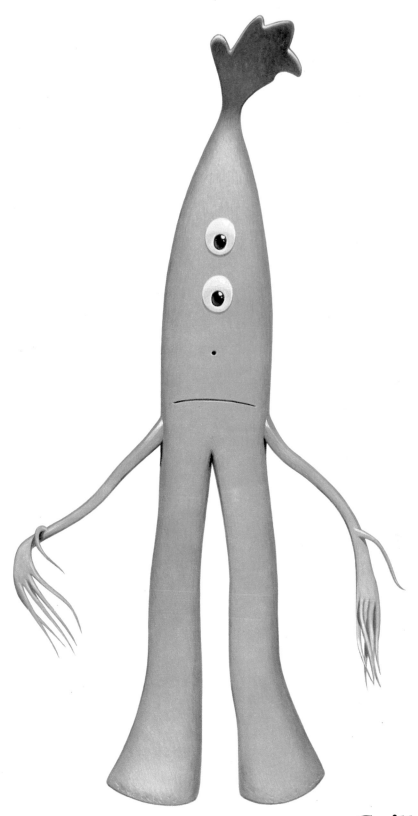

Czill

DEMON

Source:
A Plague of Demons
Keith Laumer

Physical Characteristics:
Demons are lean, somewhat doglike beings, two meters in length. The Demon's pale pinkish-gray body is sparsely covered with short, stiff bristles. Thin fore and hind legs end in pale handlike paws. The yellowish-white skull-like head has needle-sharp teeth, a ragged black tongue, and glowing red eyes. Demons are ferociously strong, and can move on either two or four feet at a swift but awkward gait. Their Earth weight exceeds 140 kilograms. Demons are oxygen breathing beings with copper-based circulatory systems. Flesh tissue is almost crystalline, and the hair contains metallic fibers.

History:
The Demons are the primary servants of a vast intergalactic intelligence engaged in an unimaginably ancient total war against all entities with creativity and free will. Ruthlessly logical, the Demons direct an army of human intelligences imprisoned in robot bodies. A powerful telepathic hypnosis enables them to induce humans to see them as members of their own kind.

Habitat:
The Demon planet of origin is unknown.

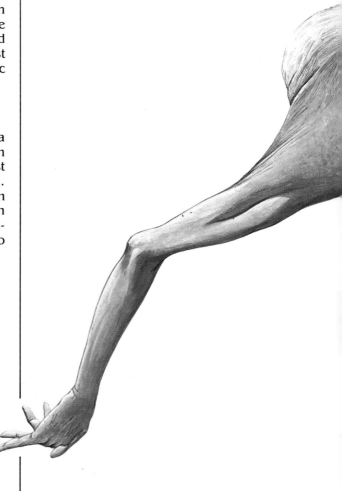

Demon

DEMU

Source:
Cage a Man
F. M. Busby

Physical Characteristics:

Demu average 1.8 meters tall. They have hard, chitinous exoskeletons that soften at the joints, to permit movement, and on the palms of the hands and soles of the feet, to enhance sensitivity. The Demu head has deep-set eyes, slightly flanged ear holes, and a toothless mouth with deeply serrated lips and a short, stumpy tongue.

Demu sexual organs are a number of small depressions and protrusions set in a ten-centimeter-wide band running up the center of the abdomen. Demu are hermaphroditic, all individuals being able to produce both eggs and sperm.

Habitat:

The Demu inhabit a twelve-planet empire, and control several more planets devoted to agriculture and populated only by slave races.

Culture:

Demu philosophy holds that the only true intelligence in the universe is Demu. When, ranging out from their own planetary system, they encountered other intelligent beings, the Demu conceived it their duty to turn these "animals" into Demu through conditioning and plastic surgery.

This philosophy dominates all the actions of the Demu, and they are unable to comprehend the resistance of other species to what they consider a generous and holy mission.

A close-up of a Demu mouth, with the tongue in a smiling position.

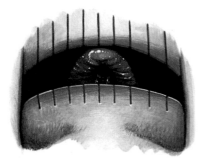

Demu

DEXTRAN

Source:
The Right Hand of Dextra
David J. Lake

Physical Characteristics:
The Dextran is a semimobile plant intelligence. Evolved from large octofungoids—carnivorous plants—the Dextran has a fifty-centimeter-long lipped opening at the top of its three-meter central sphere. Inside are several red "tongues," which trap insects and small animals and draw them into the sphere for digestion. Four thick tree-trunk-like extensions, spaced evenly around the sphere, grow horizontally out of the Dextran, which is supported about a meter off the ground by strong roots. On these trunks grow small, edible disk-shaped fruits that contain a drug that promotes telepathy. Alternating with the trunks are four large boat-shaped petals. These petals have pores that exude an attractive fragrance.

The Dextran is usually rooted deeply in the earth, but when necessary can raise itself out of the ground and walk slowly on its central root system.

Dextrans have a profound knowledge of chemistry and biology, and are able to change the form of other living things by taking them into the boat-shaped petals and processing them with their mutagenic chemicals and telekinetic powers.

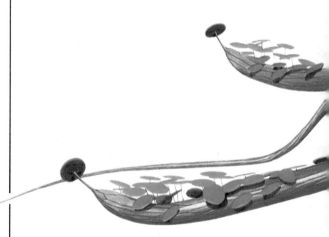

Habitat:
Dextrans are the dominant life form on the planet Dextra, which orbits the star Delta Pavonis.

Reproduction:
Only Dextran females are intelligent, the male plant having evolved into very small, totally photosynthetic forms. They reproduce by scattering spores that have been fertilized by male pollen. The Dextran can grow only in warm, moist surroundings, and the spores require many years of consistent climate to reach maturity and intelligence.

Dextran

DILBIAN

Sources:
Spacial Delivery
Spacepaw
Gordon R. Dickson

Physical Characteristics:

Dilbians are massive, fur-covered, bearlike entities with pronounced muzzles and high, intelligent brows. The male Dilbian stands up to three meters tall; the female is somewhat shorter. The thick pelt ranges in color from golden brown to reddish to black. Hands and feet have five digits each, with sharp claws. Dilbians have great physical strength, and move with remarkable agility despite their bulk, achieving speeds of up to fifty kilometers per hour at a dead run. Their deep, resonant bass voices are capable of almost deafening volume.

Dilbians have an acute sense of smell and enormous, omnivorous appetites. They are seldom clothed, often wearing no more than a satchel or a harness for an ax.

Habitat:

The planet Dilbia is an earthlike world orbiting a yellow sun. Gravity is five-sixths Earth normal. The Dilbians have not as yet achieved space travel.

Culture:

Dilbian culture has developed to a preindustrial, agricultural level with a barter economy.

Dilbians are extremely good-humored, friendly beings, given to celebration and fellowship, boasting, boistrous games, and practical jokes. They are essentially peaceful, and will take pains to avoid unnecessary conflict. Those duels or blood feuds that do occur result from a code of honor that requires the Dilbian to immediately redress any personal insult.

Fiercely individualistic, they nonetheless have a rigidly defined set of laws and codes governing their lives, administered by clan elders. Paradoxically, while Dilbians are devoted to the letter of the law, they feel it their duty to twist the spirit to the extreme. This unspoken flexibility of the code is responsible for the stability of Dilbian society.

These contrasting patches of fur from two adult Dilbian males indicate the range of pelt colors in the species.

Dilbian

DIRDIR

Source:
The Dirdir
Jack Vance

Physical Characteristics:

The Dirdir is covered with a cool, flexible skin resembling polished bone, which ranges in color from white to pale mauve. Antennae extend from the sides of the head and are incandescent, glowing brighter when the Dirdir is in the grip of strong emotion.

Dirdir move at a half hopping, half loping pace, designed for sprinting to their prey. They lack vocal cords, which accounts for their characteristic sibilant voices.

Culture:

On their home planet of Sibol, the Dirdir developed from tree-dwelling carnivores into a complex, caste-ridden race. Their virtues are based on the concept of "zest to overachieve," to prove oneself better than all others at a given thing. The Dirdir have never denied their origins as beasts of prey, and indeed believe that they must give vent to their bloodlust regularly, killing with only tooth and claw. Although they are a fierce and savage race, and their children are little better than wild beasts, there is a strong instinct for cooperation among individuals. Aided by this trait, they have developed a spacefaring culture and have left Sibol.

Reproduction:

It is known that twelve different types of sexual organs exist in Dirdir males, and fourteen types in females. Each type is compatible with one or more types of the opposite gender, and each type has certain traditional cultural attributes. While a Dirdir's basic gender is obvious from the being's skin color and body size, its exact type of sexual organs is a Dirdir's most closely-guarded secret. No outsider has ever learned the complex social conventions that surround and restrict Dirdir reproduction.

A three-toed Dirdir foot with claws unsheathed. These claws make deadly weapons.

Dirdir

GARNISHEE

Source:
Star Smashers of the Galaxy Rangers
Harry Harrison

Physical Characteristics:

The Garnishee are tree-shaped entities with four trunklike legs, over one meter in height from their feet to the long tentacles at the top of their bodies. These tentacles are used for grasping objects. The torso is studded with small pitted openings, probably sense organs. A ring of twenty-three eyes encircles the Garnishee body at mid-trunk; above the ring of eyes is a large mouth opening. The brain is located in one of the feet.

The Garnishee are omnivorous, with a diet strikingly similar to that of humans, except that they eat glass bottles for their intoxicating effect. They exude an extremely noxious body odor.

Habitat:

The Garnishee inhabit the planet Dormite, orbiting the binary star pair Alpha and Proxima Centauri.

Culture:

This ancient and highly advanced race has been locked in a merciless struggle with the dread mental leeches, the Lortonoi, and their mind-slaves, the Ormoloo, for over 10,000 years. The Garnishee have lived for centuries in underground bunkers filled with labyrinthine tunnels.

Garnishee worship the god Cacodyl; nothing else is known of their religious practices.

A cutaway view of the Garnishee brain, located in one of its four feet.

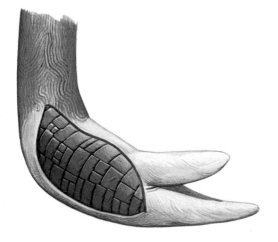

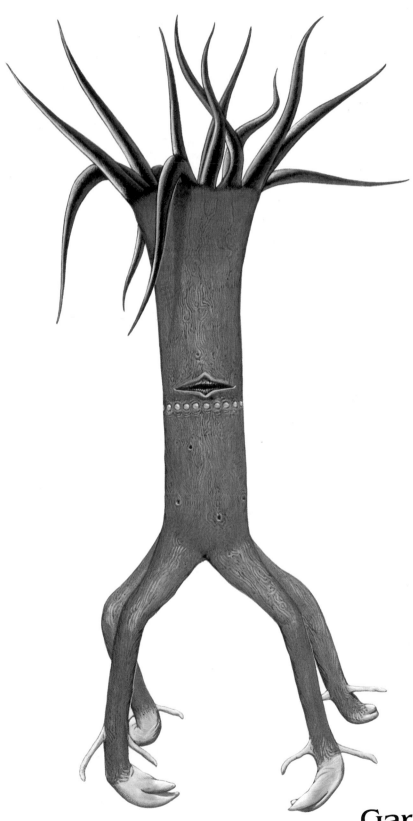

Garnishee

GOWACHIN

Source:
The Dosadi Experiment
Frank Herbert

These two Gowachin eyelid tattoos indicate membership in a particular phylum. The tattoo on the left represents Running Phylum; that on the right, Deep Swimmers' Phylum.

The fighting claws of the Gowachin in the retracted and extended positions.

These are characteristic fertility marks, found on the arms of a Gowachin female.

Physical Characteristics:

The Gowachin are amphibious, froglike beings, about two meters tall. Their skin is a smooth, wet, pale green, with webbing between the three long fingers and toes. The fingers have four-centimeter-long fighting claws that are retractable. The Gowachin have lungs, used when on land, but they also have greenish-white gill-like ventricles on the chest, which are used to extract oxygen from water. When a Gowachin is out of water, the ventricles tend to open and close rhythmically and involuntarily.

The Gowachin has no rib cage, making its torso flat and narrow. The heart is located in the lower abdominal area.

Habitat:

The Gowachin live on many planets, but their home world is called Tandaloor. They live as much in water as out of it, with waterways connecting their buildings and serving as passageways within their homes.

Culture:

Gowachin society is structured around large extended families, or *phylums*, each of which owns a huge family hall. A Gowachin's phylum affiliation is indicated by tattoos on its eyelids. Most members of a phylum are born into it, but it is possible for outsiders to make an application to join a new phylum.

Gowachin believe that their females have little intelligence, and no function beyond breeding. The females are sequestered in the *graluz*, or breeding pool, deep inside the Gowachins' house. The graluz is accessible only through underwater tunnels. There, the young *tads* are hatched. It is also in the graluz and tunnels that the adult males subject the tads to the weeding frenzy, chasing and killing all those not swift and strong enough to escape.

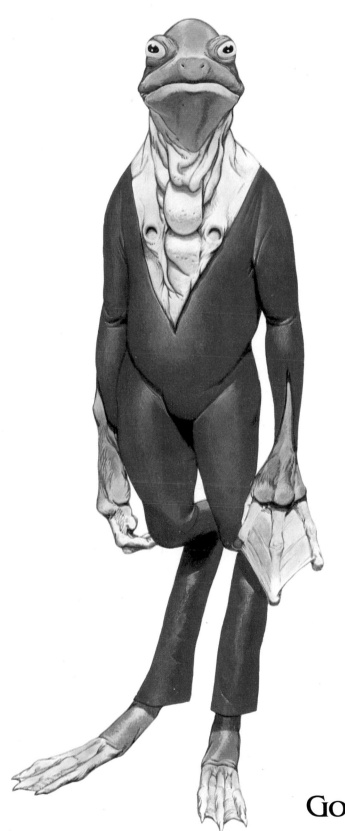

Gowachin

GUILD STEERSMAN_____

Source:
Dune Messiah
Frank Herbert

Physical Characteristics:

The Guild Steersmen are pale-skinned humanoids over two meters in length, with large webbed feet and hands. They have tiny, dark eyes and thin mouths.

The Guildsmen live in a gaseous medium imbued with a hallucinogenic spice, *melange*, which heightens their innate powers of prescience. The addictive spice comprises their entire diet; Guildsmen eat, drink, and breathe melange. Their daily life cycle shifts from a "night" period of extreme torpor to a spice-induced "day" of activity.

The distinctive Guild uniform is normally fitted with a utility belt containing the Steersman's spice, devices, and weaponry.

Habitat:

The Guild Steersmen's planet of origin is lost to history. They are now born, live and ultimately die in the weightless environment of deep space. When they must make planetfall, they survive in antigravity tanks.

Culture:

As prime navigators for a vast intergalactic empire, the Guildsmen play a strategic political role in the balance of power. Their foresight enables them to guide a Guild spacecraft at translight speed into areas of least probable danger. This prescience makes Guildsmen natural conspirators, since they can predict the activities of their opponents and the success or failure of any action.

The Guild Steersmen are cynical, power-conscious beings with no allegiance to any government. They use their powers and the strength of their Guild to preserve their role in Imperial affairs and to safeguard their supply of melange.

The flared nostril of a Guild Steersman reveals delicate gill filaments, saturated with the addictive spice melange.

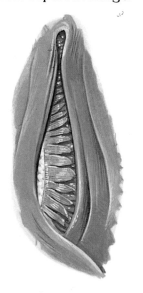

Guild Steersman

ISHTARIAN_____

Source:
Fire Time
Poul Anderson

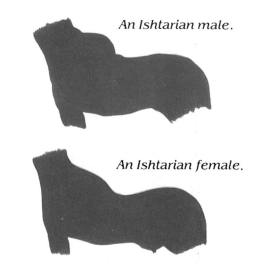

An Ishtarian male.

An Ishtarian female.

Note the differences in body size and shape.

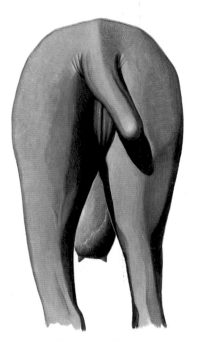

A rear view of a female Ishtarian, showing her brightly colored external genitalia and udders.

Physical Characteristics:

The Ishtarian, with its leonine body and nearly human torso, stands about two meters tall. The body is covered with a mosslike plant, leafy on head and mane, which lives in symbiosis with the Ishtarian, removing carbon dioxide and wastes from the being's bloodstream and returning oxygen and vital minerals. In addition to providing the Ishtarian with a more efficient metabolism, the symbiotic plant acts as a last-resort food supply for the omnivorous entity.

Skin color among the Ishtarians varies widely, from very light brown to nearly black. Females are generally more slightly built than males, and their hearing is even more acute than that of their sharp-eared consorts. Ishtarians live long lives, ranging from 300 to 500 years.

Habitat:

The planet Ishtar circles the star Bel. Bel is a member of the three-sun system of Anubelea, and once every thousand years, Ishtar's orbit brings it very near one of Bel's companion stars. The resulting increase in stellar radiation causes a terrible rise in the temperature of Ishtar, bringing drought, famine, and finally burning off most of the vegetation. The Ishtarians call this *Fire Time.*

Culture:

Long life-spans, coupled with the thousand-year cycles of building and destruction caused by the Fire Time, have left the Ishtarians at a comparatively low level of technology, despite their inventive and flexible minds.

The inhabitants of the northern continent are nomadic herders and hunters. Further south, where the Fire Time is less destructive, the culture is basically agricultural, with craftsmen gathering into villages on trade routes. On the numerous island chains in the southern waters, the Ishtarians have developed shipbuilding into a fine art, and handle most of the coastal trading.

Ishtarian

IXCHEL

Source:
A Wrinkle in Time
Madeleine L'Engle

Physical Characteristics:

The Ixchel are compassionate, gentle entities. Their tall, elegant, muscular bodies are covered with short, silky gray hair. Graceful tentacles extend from each of their four powerful arms; these tentacles act both as fingers and as speech organs. Other softly waving tentacles on the head function as receptors of sound and thought. The Ixchel have no eyes, and cannot understand what visually oriented beings describe as sight. The Ixchel exude a delicate fragrance that is soothing and attractive, and the gentle touch of their tentacles can ease pain and heal.

Habitat:

The planet Ixchel is earthlike, but because the atmosphere is opaque it receives only diffuse gray light during the day, and at night is completely dark. The vegetation of the planet has not developed bright colorations: the predominant hues are brown and gray. The Ixchel live in great stone halls adorned with monumental carvings.

Culture:

Very little is known about the culture and society of the Ixchel. They are at war with a totalitarian group mind that inhabits another planet in their system, and that plans to eventually absorb the entire inhabited galaxy.

The Ixchel make glorious music with their tentacle voices. They have a deep understanding of mathematics, and are able to travel through space and time via the power of their minds. With their gentle, compassionate natures and their deep core of emotional strength, they provide support for all the other beings engaged in the war.

A high magnification of the downy hairs that completely cover the Ixchel's muscular frame.

50

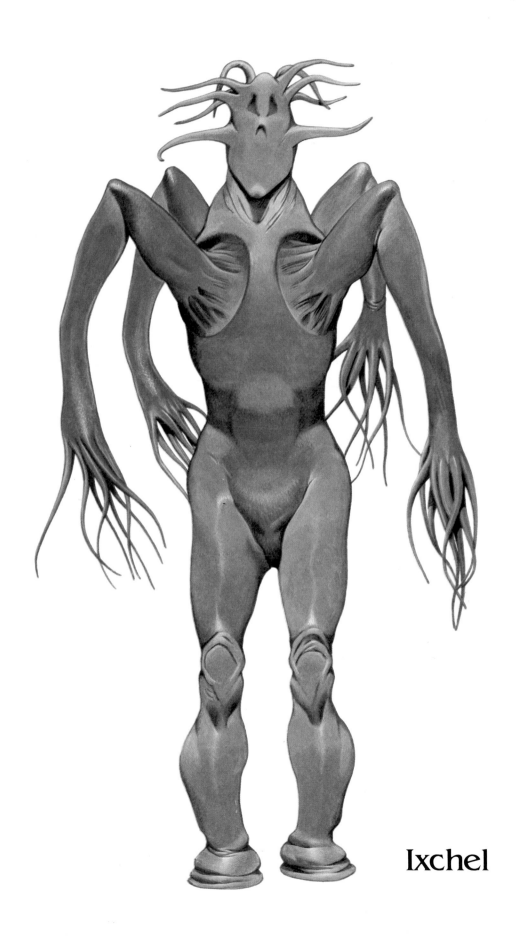

Ixchel

IXTL _____

Source:
The Voyage of the Space Beagle
A. E. van Vogt

Physical Characteristics:

The Ixtl has a long, cylindrical body with four arms and four legs. The limbs terminate in hands and feet with eight long, wirelike fingers. The body is an even, metallic red. The Ixtl's head is round, set on a short, thick neck. It has a long, gashlike mouth and two red, glowing eyes. The being has complete control over a vast web of force emanating from its body, as well as over the atoms of its body. It can change its structure and solidity at will, passing through solid matter as easily as through air. The Ixtl feeds on energy, and drops to a lower level of life force when deprived of sufficient amounts. The Ixtl can survive in any environment, even the depths of intergalactic space, for unlimited amounts of time. It can be killed only by a force such as an atomic explosion, which suddenly and completely dissolves the binding energy of the atoms of its body.

Habitat:

The Ixtl evolved on the planet Glor, and made it the center of an interstellar empire. Billions of years ago, Glor was destroyed by massive atomic attack, and only one of the Ixtl is known to have survived.

Reproduction:

The Ixtl reproduce by laying eggs in the body cavity of a living host. Within six hours of implantation the eggs hatch, and the young Ixtl eat their way out of the host body. The nourishment from this flesh allows the young to grow until they develop their force field and can absorb energy directly. These eggs can live dormant for long periods of time until an appropriate host is found. The Ixtl encountered by a Terran ship had been carrying six eggs for billions of years, and they proved viable when introduced into a human host.

A parasitic egg, three hours old; this egg was removed from the body of a young Terran host before it could hatch.

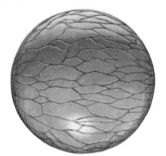

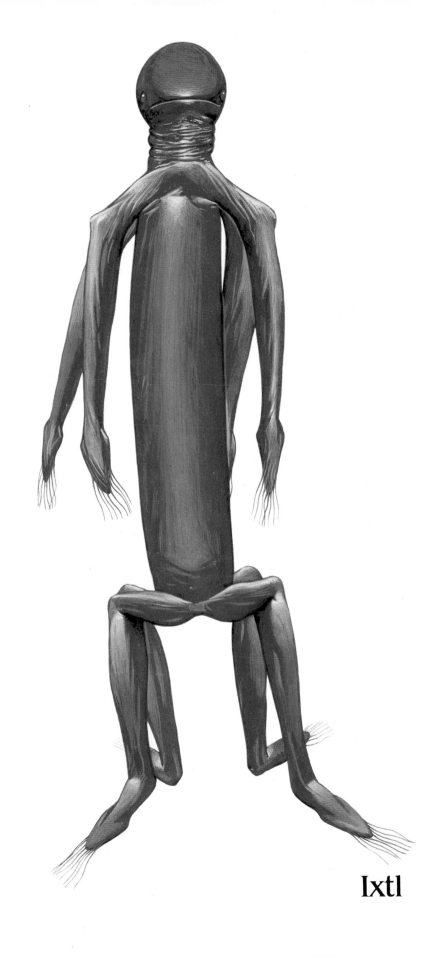

Ixtl

LITHIAN

Source:
A Case of Conscience
James Blish

Physical Characteristics:

Lithians are bipedal reptilian entities, up to four meters in height. The eyes are covered by both outer eyelids and nictitating membranes. Recessed breathing pits are located on the upper jaw. Below the lower jaw are membranous wattles. Both the wattles and a vestigial crest on the crown of the skull change color in response to emotion, ranging from pale blue to deep violet.

Habitat:

Lithia, the second planet of the star Alpha Arietes—some fifty light-years from Earth—has a gravity of .83 Earth normal, and a much lower percentage of heavy elements such as iron, lead, and gold. Much of the planet's three large land masses are covered in dense jungle.

Life Cycle:

The Lithian life cycle precisely recapitulates its evolution. Females bear eggs in marsupiallike pouches. At hatching, eggs are jettisoned into shallow sea water. The young then progress through orderly stages as fish, amphibian, and a kangaroolike lower mammal, finally evolving into mature adults.

Culture:

The Lithians have developed a high level of technology despite the lack of heavy metals. Among the scientific fields in which Lithians equal or surpass humans are electronics and electrostatics, chemistry, optics, and ceramics.

Lithians have a sophisticated ethical code, strikingly similar to Christianity but completely devoid of religious content. Their cooperative, altruistic society is based entirely on reason. There is no crime, warfare, or deviant behavior.

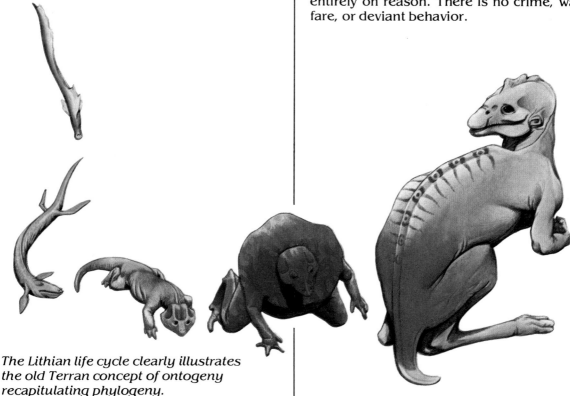

The Lithian life cycle clearly illustrates the old Terran concept of ontogeny recapitulating phylogeny.

Lithian

MASTER

Sources:
The White Mountains
The City of Gold and Lead
The Pool of Fire
John Christopher

Two Master complexion tone variations. As far as is known, skin color is not a basis for discrimination in Master culture.

Physical Characteristics:

The Masters have thick, conical bodies, three to four meters in height, tapering from about 1.5 meters to thirty centimeters in circumference. The body is supported on three short legs, and three ropelike tentacles extend from mid-body. Dry, smooth-textured skin ranges in color from pale green to dark greenish brown.

Three eyes are arranged in a triangle above two orifices; the upper for speaking and breathing, the lower for ingesting food and water. The Masters are very long-lived.

Extremely strong, the Masters move in a rhythmic, rolling gait, but can achieve much higher speeds with a rapid whirling motion. Their speech is a series of squeals and gutteral noises.

Habitat:

Although the Masters' home world is not known, its attributes may be surmised from the artificial environments the Masters create for themselves on colony worlds. Artificial gravity is kept at at least twice Earth normal. The atmosphere is a hot (about 40°C) moist greenish gas, perhaps chlorine based.

Culture:

The Masters are a conquering race from a distant star system who have enslaved the inhabitants of colony worlds with hypnotic control.

The number three predominates in their culture. The Masters move about in tall tripod machines, their buildings are pyramidal, windows and other structures are triangular, and time is divided into nine intervals.

Although little is known about their social structure, it appears that the Masters are solitary beings with an elaborate caste system. Their few communal activities include social dancing and an intricate game called the Sphere Chase.

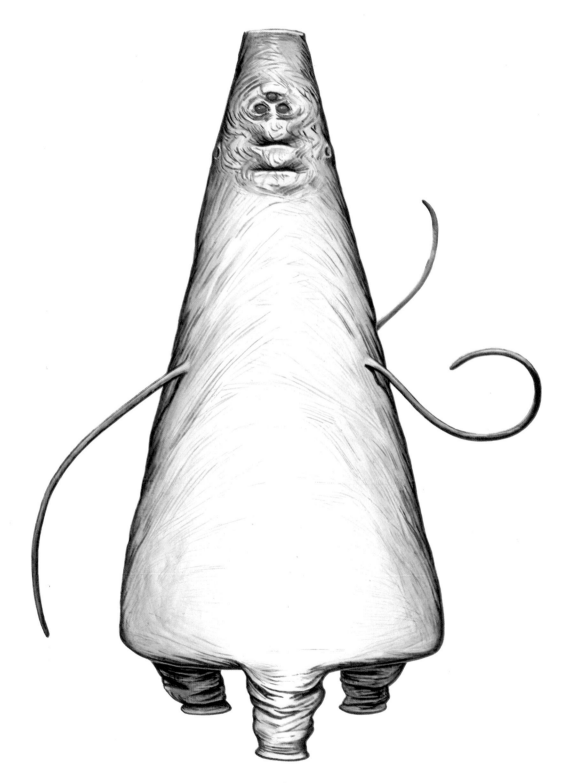

Master

MEDUSAN

Source:
The Legion of Space
Jack Williamson

One of the Medusan's four eyes. Each is approximately twelve inches long.

A detail of the flotation organ located on the underside of the Medusan torso. There is no satisfactory explanation as to how these organs function at the present time.

Physical Characteristics:

The Medusan is a translucent green hemispherical entity six meters in diameter, with fifteen-meter-long tentacles descending from the lower rim. These tentacles are black, about two centimeters in diameter, and very strong. Four large purple eyes, spaced evenly around the body, are surrounded with thin black ragged membranes. These eyes have no mechanism for controlling the amount of light that enters, and seem to be luminescent.

In the center of the bottom surface is a circular, glowing organ, one meter in diameter, which produces the forces that permit the Medusan to float and maneuver in the air. Its method of doing so is unknown.

The Medusan cannot hear sound vibrations, and communicates by generating radio pulses on the shortwave bands.

Habitat:

The Medusans are an ancient race, living on the single planet that orbits Barnard's star. This planet, 38,000 kilometers in diameter, rotates slowly under the red light of its sun. The atmosphere contains oxygen and a high percentage of helium. The Medusans occupy a single huge city on their world, from which they launch invasions of other planets to obtain fuels and metals.

Medusan

MERSEIAN

Source:
Ensign Flandry
Poul Anderson

The resting and locomotive postures
of an adult Merseian.

A close-up of the fine scales surrounding
the Merseian ear.

Physical Characteristics:

The Merseian is a two-meter-tall entity of immense strength. Merseians are mammalian—they bear their young alive and suckle them—but they very clearly resemble their reptilian ancestors. Their pale green skin is faintly scaled. The strong, heavy tail is used both as a back prop when the Merseian is at rest and as a weapon in close-quarters fighting.

A row of triangular barbs runs from over the heavy brow ridges to the end of the tail. The Merseian's eyes are small and jet black.

Habitat:

The Merseians are oxygen-breathing beings who must live on planets with plenty of free water and moderate temperatures, resembling their home world, Merseia.

Culture:

When the Merseians were first discovered by Terran traders, they occupied one planet, and had not developed space travel. Their society was feudal, with a tradition of loyalty and responsibility between the nobility and their client families. After first contact, the Merseians acquired interstellar spacecraft, and began building their own empire. They retained their class structure: the Emperor is elected by a governing council of noble clan heads. The Merseians are militaristic, proud of displaying their courage in battle. They have a tradition of independence that makes them impatient with bureaucracy.

Merseian

MESKLINITE

Source:
Mission of Gravity
Hal Clement

Physical Characteristics:

The Mesklinite is thirty-five to forty centimeters long and five centimeters in diameter, with eighteen pairs of legs. Each of the legs ends in a suckerlike foot, enabling the Mesklinite to tightly grip any surface. Forward pincers function as hands, while a rear set is used for anchoring the Mesklinite in position when necessary. The being's four eyes surround a mandiblelike mouth.

The Mesklinite has no lungs, absorbing hydrogen directly from the air into its body through pores. It has a complex circulatory system, with a heart located in each body segment. Alongside the digestive tract is an internal siphon system, originally used by the Mesklinite's remote ancestors for submerged propulsion, and now used to produce speech. The Mesklinite's voice has a wide frequency range, from very low to what humans consider ultrasonic, and can be extremely loud.

The Mesklinite's body is chitinous, with a heavily armored upper surface. Its pincers and mandibles are of a tough and very sharp material, and it is tremendously strong.

Habitat:

The planet Mesklin circles a double sun, Belne and Esstes, requiring 4.8 earth years for a complete orbit. Mesklin makes a complete rotation on its axis once every eighteen minutes. The tremendous forces generated by this rate of rotation have flattened Mesklin into a lozenge shape, 76,800 kilometers in diameter at the equator and slightly less than 32,000 kilometers along the axis from pole to pole.

Because of Mesklin's unusual shape, the gravitational force on the planet varies widely between the poles and the equator, ranging from three to seven hundred times Earth normal. The temperature on Mesklin can rise as high as $-140°C$ in the summer, melting off the ammonia snow and evaporating the methane oceans.

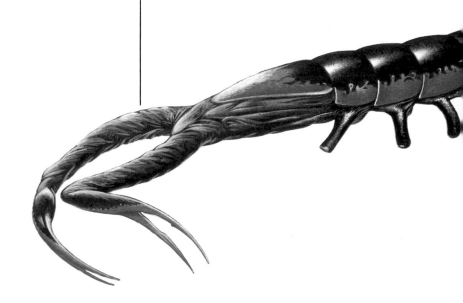

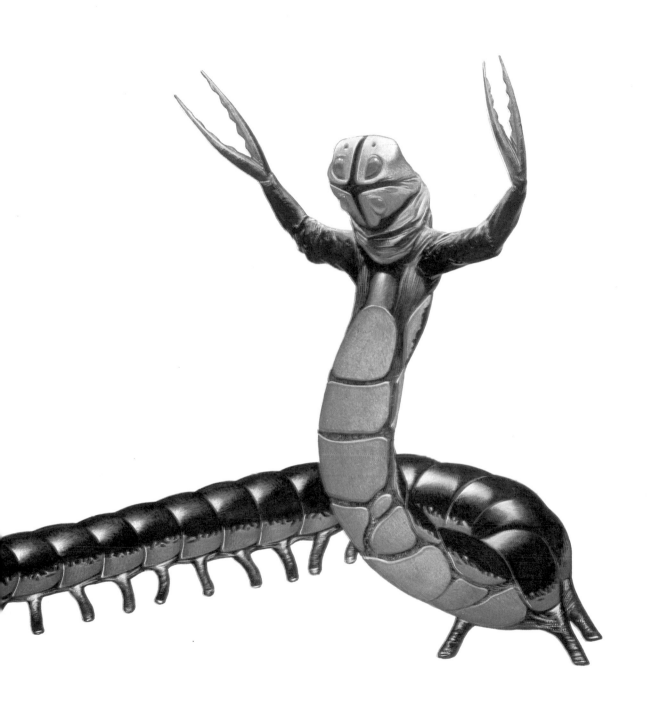

Mesklinite

MOTHER

Source:
Strange Relations
Philip José Farmer

Physical Characteristics:

The Mother is a large, intelligent organism with protective outer camouflage resembling a large boulder in size and texture. Within the outer shell is a warm, fleshlike skin. A long, stalklike antenna is capable of broadcasting messages in radio frequencies and receiving information from other members of the species.

A great cavity in the Mother's side, edged with a hairlike fringe, opens for feeding or mating. The interior of the being is an egg-shaped chamber approximately nine meters in length. The smooth inner walls are moist and reddish-gray in color, and covered with small red and blue tubes that may function as veins and arteries. On the walls are groups of long tentacles that can reach out through the opening.

The Mother has several hearts, four stomachs and powerful lungs, and scent glands used to lure prey. A deep mouth cavity lined with thousands of razor-sharp fangs opens from the inner chamber. The Mother is omnivorous, eating both the flesh of her prey and vegetable matter.

Habitat:

The Mothers inhabit the planet Baudelaire, a cloud-covered, earthlike world orbiting a double sun.

Culture:

Mother society is an elaborate hierarchy with status based on age, size, and power of broadcast. A "Queen" Mother dominates the radio frequency broadcasts, and inferior members may only broadcast with her permission.

Reproduction:

All Mothers are immobile and female. All mobile beings are therefore considered male. The Mother seduces her mates by exuding an attracting musk. The "mate" is then brought inside the Mother and induced to attack a *conception spot*—a large, circular swelling on the inner wall. The abrasion by beak, claw, or fang starts the conception process. The mate is then devoured by the mouth cavity.

After conception, the spot swells into a bag, in which develop ten young. The young are dropped into the womblike interior of the Mother, where they are nurtured from the organic stew in her stomach. When they are over a meter in length, the young are pushed out of the womb. They roll until they reach the empty shell of a dead adult or the top of a hill. Here they put out thin, threadlike tentacles that probe deep into the soil for water and nutrients. Using available materials, they fashion layers of protective shell to protect themselves from predators.

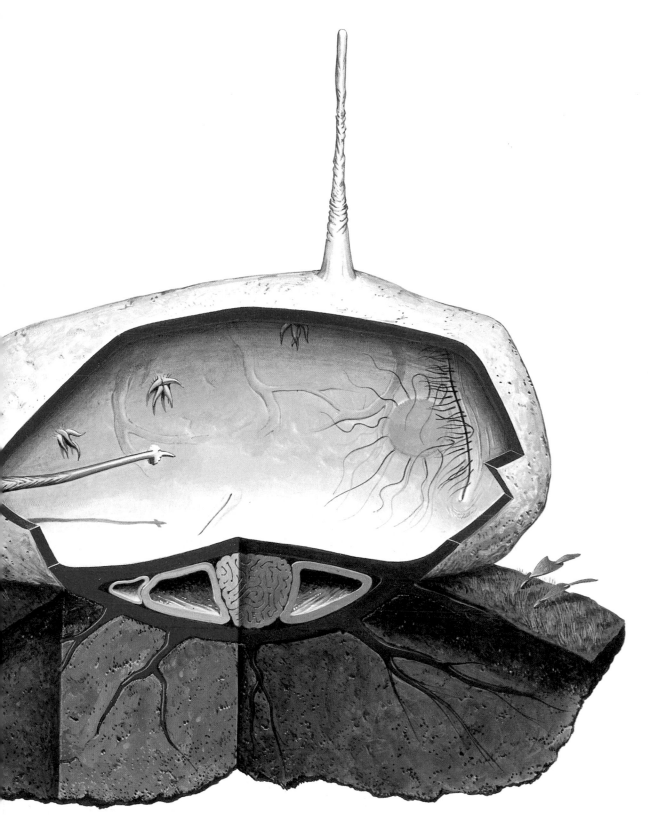

Mother

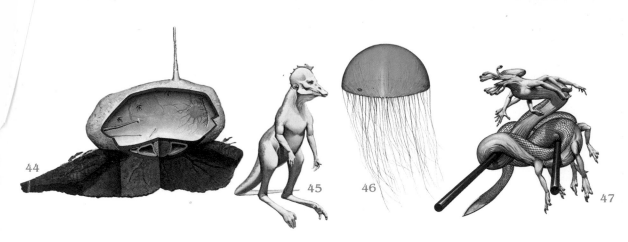

COMPARATIVE SIZE CHART

KEY	SIZE		KEY	SIZE	
1. Old Galactic	.45	m	26. Cryer	2.0	m
2. Radiate	up to 1.0	m	27. Regul	1.8	m
3. Mesklinite	.35	m	28. Sirian	2.4	m
4. Athshean	1.0	m	29. Pnume	2.0	m
5. Triped	1.2	m	30. Wayne D.		
6. Cinruss	1.5	m	Barlowe	1.8	m
7. Puppeteer	1.25	m	31. Salaman	1.7	m
8. Abyormenite	1.2	m	32. Ixtl	3.4	m
9. Riim	1.5	m	33. Czill	2.0	m
10. Overlord	3.6	m	34. Gowachin	2.0	m
11. The Thing	1.5	m	35. Dirdir	2.0	m
12. Thrint	1.25	m	36. Guild		
13. Vegan	1.2	m	Steersman	2.2	m
14. Uchjinian	2.0	m	37. Tran	2.0	m
15. Tyreean	1.8	m	38. Old One	3.8	m
16. Ruml	1.5	m	39. Merseian	2.0	m
17. Cygnan	1.5	m	40. Sulidor	3.0	m
18. Chulpex	1.8	m	41. Master	3.5	m
19. Demu	1.8	m	42. Dilbian	3.0	m
20. Garnishee	1.5	m	43. Ixchel	3.2	m
21. Polarian	1.8	m	44. Mother	9.0	m*
22. Demon	2.0	m	45. Lithian	4.0	m*
23. Slash	1.0	m	46. Medusan	15.0	m*
24. Cygnostik	2.0	m	47. Velantian	10.0	m*
25. Ishtarian	2.0	m	48. Dextran	3.8	m*

NOTE: Extraterrestrials are shown in scale wherever possible. Exact sizes are not always mentioned in the source; therefore, we had to make guesses based on their descriptions. Some aliens are painted in perspective and may appear smaller or larger than the height indicated above. Other entities would be too large to show in relative proportion. These aliens are presented out-of-scale and are indicated by asterisks (*).

OLD GALACTIC

Source:
Legacy
James H. Schmitz

A diagram of an Old Galactic in motion. Its flexible body extends to cover distances. However, Old Galactics generally prefer to travel in synthetic host bodies.

Physical Characteristics:

Old Galactics are about forty-five centimeters long, and mass about eight kilograms on Terra. Their bodies are dark green marbled with pink streaks. The Old Galactic's surface is smooth and warm, and can be very pleasant to the touch if contact is desired. But they are able to store an electric charge and can release that charge at will, badly shocking any entity that handles them against their wishes.

The Old Galactics are essentially nervous systems, with no visible sensory organs of their own. They are able to insinuate themselves into the bodies and nervous systems of other, more mobile beings, and force the beings to take any action required by the Old Galactics.

The Old Galactics perceive time much more slowly than most intelligent beings, and live correspondingly longer lives.

Culture:

The Old Galactics have a deep understanding of biology and genetic manipulation. Although they began their imperial expansion by taking over other intelligent beings, they soon developed an ethical bias against this. To retain their freedom of movement, they designed and built synthetic life-forms to act as specialized host bodies. Their rule over their empire is based not on brute force, but rather on their ability to manipulate genetics; a rebellious planet might find itself infected with a virulent and very selective plague until the dissident population surrendered to the Old Galactics.

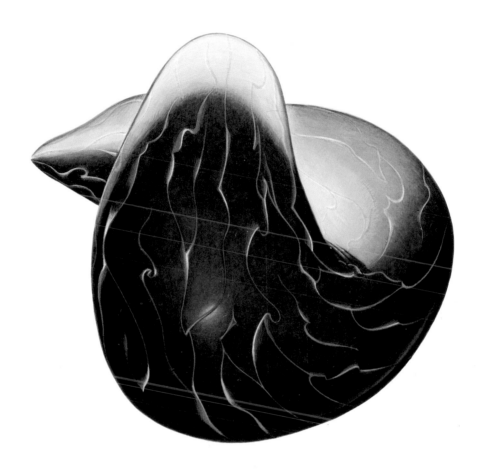

Old Galactic

OLD ONE _____

Source:
At the Mountains of Madness
H.P. Lovecraft

A detail of a closed (left) and open (right) eye. The Old Ones' lids almost totally retract from around the eyeball.

The mouth bulb closed (left) and open (right).

The prismatic cilia on the top of an Old One's head apparently supplement its vision; in the absence of light, the species is able to "see" using these organs.

Fossil footprints found in Archaean rock, made while an Old One was living on land.

Physical Characteristics:

The Old Ones are incredibly tough and durable organisms, having characteristics of both plant and animal life. They are able to live on land, breathing through an orifice on the top of their star-shaped heads, and under the sea, breathing through gills. They can also survive the rigors of space, storing air, food, and minerals and sailing the solar winds on large, membranous wings. The 1.3-meter-long lower tentacles are extremely muscular, and are used for walking and swimming.

The Old Ones' metabolism is based on carbon dioxide rather than oxygen, and their circulatory fluid is dark green. They have very complex nervous systems, with five-lobed brains and many ganglial centers. The connections between specialized structures in the nervous system and wiry cilia on the head suggest that they have senses other than the human ones of sight, smell, hearing, touch, and taste.

Habitat:

There are no living Old Ones on Earth, but their remains have been found in a huge ruined city in Antarctica. It is clear from these frozen bodies that the Old Ones lived under the most varied of conditions.

History:

Traveling through space on their great wings, the Old Ones came to Earth nearly 1,000 million years ago, seeking refuge from the overly mechanistic culture that developed in their home system. At first they lived under the sea, building cities and exploring their new world. It was then that they created the first marine life on the planet, to supply themselves with food. Later, some migrated to land, where they developed land animals and their huge beasts of burden, the Shoggoths.

During a period of great geological cataclysms, the Shoggoths acquired a rudimentary intelligence that developed until they became a threat to their creators. A millennium-long war finally caused the collapse of the Old Ones' world dominion and they retreated back to their greatest city in the Antarctic. Although they destroyed nearly all the Shoggoths, the debilitating struggle set the Old Ones on an inevitable road to degeneration and final extinction. Long before humans developed intelligence, the Old Ones were gone.

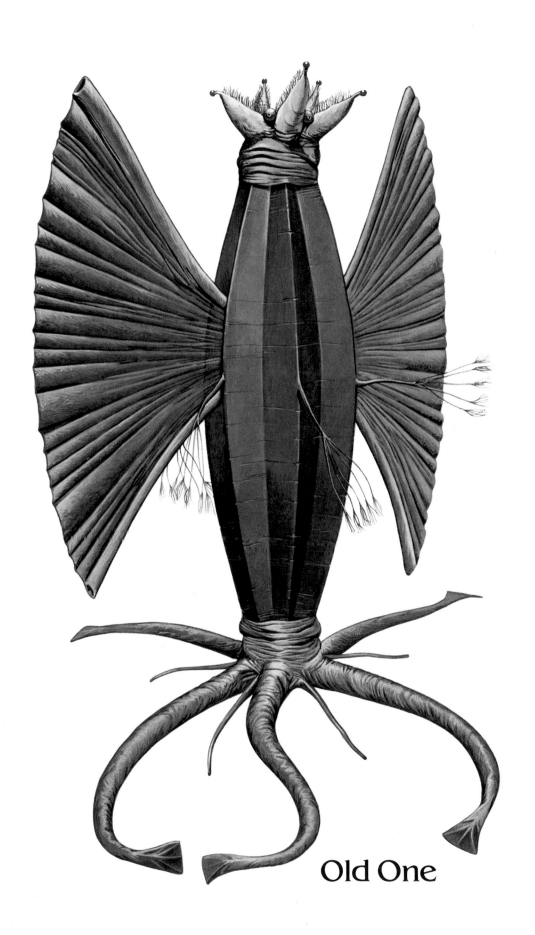

Old One

OVERLORD

Source:
Childhood's End
Arthur C. Clarke

The hand of an Overlord, showing its double opposing thumbs.

Physical Characteristics:

The Overlords are highly intelligent entities, about three meters tall. They have huge, leathery, functional wings and a long tail with a flat, paddlelike end that gives them stability in flight. Their bodies are covered with hard black armor; it is not known if they also have an internal skeleton. On the broad cheeks are twin breathing orifices lined with tiny white hairs, which can be seen to flutter as the being inhales and exhales oxygen. The Overlords' eyes are very sensitive to bright light, and they prefer the dimmer red end of the spectrum. The hands have five fingers and two thumbs, one on either side.

The Overlords do not require sleep, and are able to learn and understand an incredible amount of information in a short time.

Habitat:

The Overlords presently inhabit a large planet orbiting the star NGS 549672, a red sun. They originated on a much smaller world, with a thick atmosphere and very low gravity. The cities of the Overlords are designed for beings that can fly; there are few if any walkways, and doors are placed at any convenient elevation.

Culture:

The Overlords have developed a highly technological and scientific culture, with a star-drive that enables their spaceships to achieve 99 percent of the speed of light. This, in addition to their long life spans, has given them the freedom of the galaxy, enabling them to study and collect specimens of all life forms. Their planet is organized into about a thousand small cities, each of which is devoted to the study of a single branch of knowledge.

An Overlord wing in the extended position.

Overlord

PNUME

Source:
The Pnume
Jack Vance

The Pnume's flexible toes allow it to grasp objects with its feet.

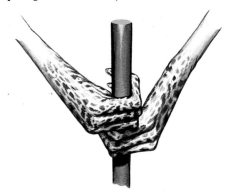

The specialized rasping mouthparts of the Pnume look almost mechanical in close-up.

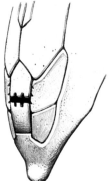

Physical Characteristics:

The Pnume is two meters tall, thin, and covered with a hard, pale exoskeleton that softens to flexibility at joints, fingers, and toes. The head is long, with complex rasping mandibles concealing the mouth. The eyes are shadowed in deep sockets. The long, strong arms end in hands with two bony fingers. The feet have three strong articulated toes, and are mottled red and black.

The Pnume's legs are jointed so that the knee bends backward. Shoulders and hips are flexible ball joints and the neck is hinged. By bending over backward, swinging the arms back, and releasing the neck hinge, the Pnume becomes quadrupedal. It can move quickly and easily in this fashion.

Habitat:

The Pnume live in vast underground caverns and passages that honeycomb the planet Tschai. Their underground world is decorated with polished, glowing crystals and mosaics, and has numerous secret passages and access to dozens of broad, swift rivers.

Culture:

The planet Tschai has been invaded at least ten times in the seven million years of recorded Pnume history. The Pnume take the position that all the invasions and wars on the surface of Tschai are grand spectacles produced for their education and amusement. Dressed in their dark cloaks and broadbrimmed hats, they appear secretly at night on the surface and collect artifacts and specimens of the invading races. These items are cataloged and stored in the Pnume museums of Foreverness.

Pnume

POLARIAN

Source:
Cluster
Piers Anthony

Physical Characteristics:
The Polarian is a teardrop-shaped entity, approximately 1.8 meters tall when fully extended. A muscular socket at the bottom of the body holds a large wheel. At the upper end, the body tapers into a flexible tentacle (called a *trunk* if the Polarian is male, and a *tail* if female), which terminates in a smaller ball, held in place by a similar socket. Both

The final embrace of a Polarian couple, in which the male selflessly gives the female his own ball.

In mating, the male Polarian releases his own wheel and takes the lower half of the female's wheel into his empty wheelchamber.

ball and wheel are spun in their sockets by powerful, adhesive muscles: spinning the wheel, a Polarian is capable of moving at speeds approaching 112 kilometers per hour and of rapidly changing direction. By spinning the smaller ball against any hard surface, the Polarian can produce a wide range of sounds, and is capable of approximating human speech.

The Polarian's soft and flexible body has no bones. The dark brown skin is smooth and waxy; it acts as a radiation receptor, transmitting sensory information to the Polarian. The Polarian's skin can also radiate light: its glow reflects the Polarian's emotional state.

Habitat:
Polarians are carbon/oxygen-based beings, able to live, like humans, in a wide range of climates. They have developed interstellar travel, and occupy a sphere of influence 400 light years in diameter, centered around the star Polaris in their home system.

Reproduction:
The Polarian reproduces in an unusual fashion. The male and female circle each other, the male following a seductive scent trail laid down by the female's wheel. They circle faster and faster, spiraling in toward each other until they meet. The male entwines his trunk around the female's tail, and their two balls touch in an electrifying spinning kiss. Slowing, the male releases his wheel, and takes the exposed lower part of the female's wheel into his now vacant wheelchamber. Their sockets seal tightly against each other, and together they spin the wheel, generating intense heat and releasing special enzymes. There is a moment of rapture as an electrochemical shift occurs, sending a shock through their bodies. The mutual wheel then rolls free of the lovers, steaming and soft. In moments it unfolds, stage by stage, into a fully developed young Polarian, able to care for itself.

The male retrieves his wheel, while the female transfers her ball into her wheel socket, where it will gradually grow into a full-sized wheel. In a final embrace, the male gives her his ball to replace her now missing communication organ, rendering himself mute until he can grow a new one. The two part, never to mate again.

Polarian

PUPPETEER

Sources:
Neutron Star
Ringworld
Larry Niven

The mouths of the Puppeteer function as hands; the knobs illustrated are its "fingers."

When frightened, a Puppeteer curls up into a little ball.

Physical Characteristics:
The Puppeteer is a 1.25-meter-tall herbivorous entity with two flat, brainless heads at the ends of long, sinuous necks. Each head has a single eye, a mouth with square, heavy teeth, and a forked tongue. The mouths of a Puppeteer function as its hands, aided by sensitive fingerlike knobs. The Puppeteer's brain is located in a bony hump between the two necks. This hump is covered with a thick mane of hair, which ranges in color from golden to dark brown. The rest of its graceful body is covered with suedelike creamy white skin.

The Puppeteer's remarkable brain enables it to carry on two conversations at once, using the two heads independently. Its speaking voice is usually contralto, but when singing, a Puppeteer can produce a full orchestral sound.

Habitat:
The location of the Puppeteer's home planet is one of the most closely guarded secrets in the universe.

Culture:
The Puppeteers are a very old species, and have had a highly developed technological civilization for millions of years. They are an extraordinarily cowardly race: they will go to any extent to avoid physical danger. This characteristic, coupled with their high level of intelligence, has enabled them to develop the safest and most reliable devices in the universe. Most of their contact with other intelligent beings is through the General Products Company, a Puppeteer-owned concern that manufactures and sells completely indestructible spaceship hulls, as well as life support equipment for interstellar travel. Puppeteers are known throughout the universe as shrewd businessmen whose favorite method of negotiating is through blackmail.

Puppeteer

RADIATE

Source:
Memoirs of a Spacewoman
Naomi Mitchison

Physical Characteristics:

The Radiates, who live on a planet designated Lambda 771, have evolved from a radial life form somewhat like a terrestrial starfish. The five arms are retractable, and are studded with delicate suckers that are used for grasping tools and artifacts.

On the top of their bodies, Radiates have a ring of bright blue eyes that encircle the brain case. Their bodies range in size from a few centimeters to about one meter in diameter.

Habitat:

Nothing is known of planet Lambda 771 except that it is similar in size to Terra and has an oxygen-based atmosphere.

Culture:

Radiates live in villages composed of long, low buildings, roofed but with open sides. They decorate their ceilings with plants and fungi that grow in spiral patterns.

They do not think in terms of dualities, having instead a five-valued system of logic.

At intervals, all members of a community will join in an interlocking, wheeling dance, with those on the outer edge attempting to move closer to the center; this dance has strong emotional overtones of closeness and community unity.

Radiates wear flimsy, artificial clothing made from an unusual sticky material. When clothed, a Radiate's brain area is covered. When not in use, the clothing is retracted and stored underneath the body.

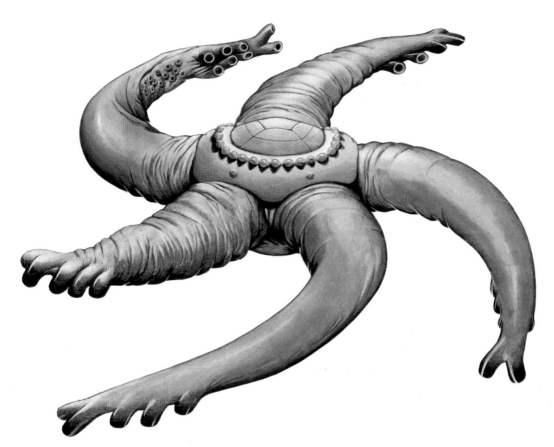

Radiate

REGUL

Source:
The Faded Sun: Kesrith
C. J. Cherryh

Physical Characteristics:

Regul are slow moving, long-lived entities whose bodies change radically as they grow older. In their youth, they are erect and relatively slender, with their bone structure visible under their gray-brown pebbly skin. Their extremely short legs are bowed, which causes the young Regul to move with a rolling gait. As the Regul ages, its body grows heavier, until nothing of the being's body shape can be determined under massive, wrinkled rolls of flesh. Soon after the Regul becomes an adult, it reaches a weight at which it is impossible for it to move itself.

Regul have two genders, but even the Regul themselves are unable to predict a youngling's future sex.

Regul possess eidetic memories; they are unable to forget anything that they see or hear during their long lives.

Culture:

The Regul have built an interstellar commercial empire, organized by clans. The immobile adults direct the activities of young Regul from their power sleds, and have the power of life and death over the younglings that owe them allegiance. As a race, the Regul are noncombative and nonviolent, except toward their own children. They prefer to hire mercenaries of other species to fight their commercial wars.

The nostrils of a Regul, either closed or flared, indicate the emotional state of the entity. These emotions cannot be translated in Terran terms.

The legs of a young Regul will virtually disappear under rolls of flesh as the being attains a respectable age.

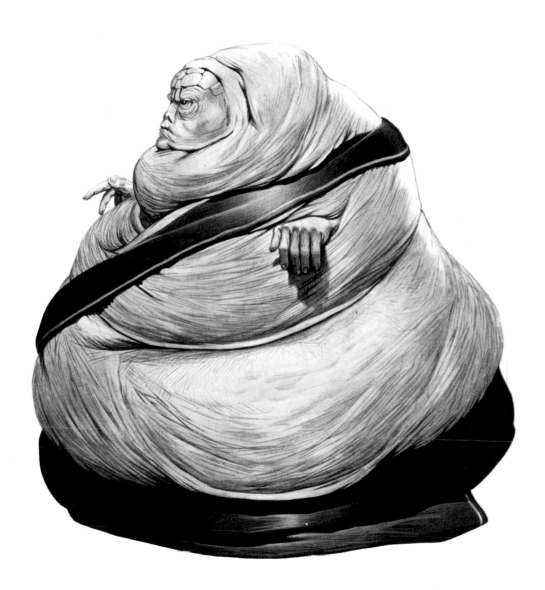

Regul

RIIM

Source:
The Voyage of the Space Beagle
A. E. van Vogt

The stunted, vestigial wings of an adult Riim.

This silhouette of the Riim head and torso illustrates the placement of the wings on the body.

Physical Characteristics:

The Riim are erect, bipedal, birdlike entities with vestigial wings. Their heads and wrists sport tufts of golden feathers. They have narrow shoulders and short arms but extremely long hands and fingers. The head has two wide-spaced eyes set over a short beak. A faint tracery of veins on the face gives the impression of nose and cheeks to a human viewer, although the Riim have neither.

The Riim reproduce parthenogenically, the offspring growing out of the body of the parent until it is mature enough to separate completely.

The Riim are telepathic, and can project illusions into the minds of nontelepathic beings.

Habitat:

The Riim inhabit an earthlike planet with slightly lower gravity. Their cities are composed of huge, dark buildings that have an internal network of catwalks and roosting platforms. Some Riim live on agricultural land, producing food for themselves and the city dwellers.

Culture:

The Riim society is in essence a telepathic group mind. Each individual is an intimate part of the race as a whole.

Riim

RUML

Source:
The Alien Way
Gordon R. Dickson

This silhouette of a female Ruml shows the distension of her marsupial pouch. Her offspring will remain in the pouch for six years.

The hinged earflap, shown in an open and closed position, protects the Ruml's sensitive inner ear.

Physical Characteristics:

Ruml are about 1.5 meters tall, and are covered with glossy fur. The narrow pelvis is tilted forward, so that a Ruml cannot stand fully erect. Their long arms, designed to be held bent at the elbows, end in hands that have eight-centimeter-long retractable claws.

Ruml are marsupials with unusually long gestation and developmental periods. Ruml are carried in the womb for three years, and after birth move immediately to the female's pouch. They remain there for six more years, developing but not increasing in size. In the sixth year, the infant Ruml begins a period of rapid growth, soon becoming too large for its pouch. Within three hours of emerging from the pouch, the Ruml is a fully functional adult, although lacking in experience and size. Over the next two years, the young Ruml achieves its full growth.

Habitat:

The Ruml occupy six planets. Their Homeworld is slightly smaller than Earth, and has barely enough water to support life. The Ruml live communally in large family palaces.

Culture:

Ruml families are large groups of related entities under the authority of one individual. Status is determined by the degree of kinship with the family head. As a result of population pressure, the Ruml have moved out from Homeworld to colonize other planets. Each Ruml who returns from a successful expedition to discover a new inhabitable planet wins the right to found a family on that planet.

The Ruml are a violent race, with a strong concept of personal honor and a custom of dueling to the death in order to defend that honor. Their loyalty is to their race as a whole rather than to any individual. Ruml are awarded medallions for achievement; their status in the community is based upon the number of awards they win. Under this system, there is great incentive to excell; but the customary penalty for failure is death.

Ruml

SALAMAN

Source:
Wildeblood's Empire
Brian M. Stableford

Salaman skin coloration is affected by climate and surroundings. Top, Salaman skin on a normal day; left, in overcast conditions; and right, in the desert. The skin below is that of a juvenile; as he grows, his skin color will darken considerably.

Physical Characteristics:

Salamen are amphibious entities about 1.7 meters tall. Young Salamen are pale blue, dappled with brown and green; the blue pigment darkens to nearly black as they mature. They shed their skins frequently.

The Salamen have a very complex life cycle, which allows them to live both on land and in the water. Eggs are laid in the water, encased in gel clusters. These eggs hatch into young aquatic Salamen, who mature as water breathers until they reach the point of metamorphosis into adults. At this time, the young Salaman may choose to change either into an adult aquatic form or a juvenile air-breathing land form. If it chooses to become a land dweller, it continues to develop until it must once again choose whether to metamorphose into an adult land dweller or to revert to the juvenile aquatic form. Once a Salaman has chosen to change into an adult form, it can change at will from adult land dweller to adult water dweller and back. At any one time, about half the Salaman population is in each form.

Culture:

The Salamen have a complex language of hand signs, used both on land and in the water. The land dwellers also use a simple vocabulary of barking and whistling sounds. In their land phase, the Salamen live primitively, with stone tools and clothing made of plant fibers, practicing simple agriculture. In their aquatic phase, the Salamen are students and teachers, engaged in protecting and educating the newly hatched.

As a species the Salamen are nonviolent and extremely social. Almost all activities are group oriented, and there is constant exchange of group members between land and sea.

Salaman

SIRIAN

Source:
The Age of the Pussyfoot
Frederik Pohl

Physical Characteristics:

The Sirian is an octopuslike tentacled being covered with bright green fur. Dozens of tiny, twinkling eyes are set into the green ruff on the Sirian's head.

Although very little is known about the Sirians apart from their appearance, it can be assumed that they do not breathe oxygen, since they must travel about Terra in complex life-support tanks. They may be three-gendered; the name of one of the Sirians on Terra translates as Alphard four Zero-zero Trimate.

Habitat:

The Sirians inhabit twelve planets orbiting the stellar pair of Sirius A and B.

Culture:

It is believed that the Sirians are an extremely warlike race. Their planetary system is heavily fortified, and they appear to mount frequent and very realistic war games. Unfortunately, the initial contact between humans and Sirians resulted in hostilities, which left the Sirian ship destroyed and only fourteen Sirians alive to be taken prisoner. These survivors have refused to give any information about their race.

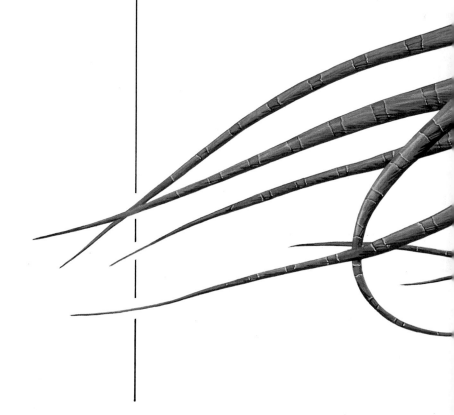

Sirian

SLASH

Source:
Kirlian Quest
Piers Anthony

Physical Characteristics:
The Slash are long, tubular entities with manipulating tentacles at either end. Sharp, metallic blades encircle their bodies, used to cut pathways, butcher food, and fight. Between the blades are laser lenses; the Slash uses the flickering light of the lasers to see and to communicate. A steady, powerful beam of laser light can flare out in attack and defense.

Habitat:
The Slash occupy a sphere of influence in the Galaxy Andromeda.

Culture:
During an early period of intergalactic wars, the Slash built a vast empire, moving onto conquered planets as slave-holding overlords. After an Andromeda alliance led by the Slash was defeated due to the efforts of a traitorous Slash, they retired to a small sphere of influence to pursue lives of tolerance and peaceful study. The Slash now suffer the distrust and prejudice of all other sapient beings.

Slash

SOFT ONE _____

Source:
The Gods Themselves
Isaac Asimov

This hardened extruded pseudopod enables a Soft One to "handle" solid objects.

Physical Characteristics:

The Soft Ones are intelligent gaseous entities held together by strong energy fields. Their bodies typically hold an ovoid shape, but they can thin or thicken their mass at will, and also extend pseudopods and "harden" them sufficiently to manipulate solid objects. They feed on sunlight, and communicate by pulsing.

Habitat:

The Soft Ones inhabit a rocky planet in a parallel universe adjacent to our own, in which differing laws of physics, such as the strength of the nuclear force, apply. We are aware of their existence through messages they have passed into our universe.

Reproduction:

The Soft Ones have three sexes: Rationals, Emotionals, and Parentals. The Rational is smooth and curved, and exults in the pursuit of reason. The Parental is more solid in form and is driven by a strong mating instinct. The Emotional is of more tenuous substance, and can "thin" its mass almost to nothing. The basic familial unit, the Triad, is made up of one of each type. The Triad reproduces when its members *melt* into one another.

At maturity, a Triad of Soft Ones is brought together with the aid of a mysterious race of beings known only as "Hard Ones." The Emotional gorges herself on sunlight to provide the tremendous energy needed for a successful melting. She then thins herself into a shimmering, colored smoke that catalyzes the melting of the Rational and Parental. At climax, the three members interpenetrate in an ecstatic union followed by a prolonged period of unconciousness. During the melting, a *seed* is formed by the Rational and passed on to the Parental for incubation. Each Triad conceives three offspring: first a Rational, then a Parental, and finally an Emotional. Shortly after the third child is grown, the Triad "passes on" in a final melting.

Soft One

SOLARIS

Source:
Solaris
Stanislaw Lem

A mimoid of Solaris in the early stages of cloud imitation.

Two rare independents, *able to actually detach themselves from Solaris's surface, creating* (left), *a birdlike form, and* (right), *a seallike form.*

Physical Characteristics:

Solaris is an immense (700 billion tons) intelligent organic entity that forms the ocean of the planet that bears its name. The planet Solaris has a diameter of about 15,000 kilometers, and orbits a binary star—a cool red sun and a hot, brilliant blue-white sun—in the Alpha region of Aquarius. While the combined gravitational stresses of the two stars should have long ago pulled Solaris apart, the world seems to have some as yet unknown method of stabilizing its orbit.

Solaris's atmosphere is completely devoid of oxygen. The wind storms, fogs, mists, and cloud formations that can instantly become viscous or solid are believed to be deliberate manifestations of the planetary mind.

The ocean's surface is studded with countless small islands, many formed out of the matter of the world ocean itself, and located predominantly in the southern hemisphere. The sea of complex organic compounds is normally a deep purple, with an undulating surface flecked with pinkish foam.

Ecology:

Out of Solaris's fluid plasma are generated an extremely diverse range of formations and an extraordinary range of textures. The transformations are believed to be based on subatomic control of the colloidal fluid by the organizing intelligence. Many of Solaris's manifestations defy conventional notions of physics on both the planetary and subatomic levels.

Although the surface of Solaris is almost infinitely variable, certain repetitive features have been identified. *Extensors* are sinuous fluid ridges many kilometers in height and length. *Mimoids* are huge structures, superficially resembling cities, which are capable of extruding masses of plasma that mimic the shapes of other things, from passing clouds to manmade objects. *Symmetriads* are complex shapes, hundreds of meters in height, which begin as glowing spheres, open into flowerlike shapes, and then form tall domes with intricate interior structures that appear to replicate mathematical equations in geometric form. Other formations are free-floating, multiwinged birdlike shapes and glowing phosphorescent bodies.

Solaris

SULIDOR

Source:
Downward to the Earth
Robert Silverberg

Physical Characteristics:
The Sulidor, towering three meters tall, is a heavy, hulking entity covered with reddish fur, with dark, deeply hooded eyes above a long, drooping nose. The Sulidoror are carnivores, with sharp, lethal teeth. They have thick, broad tails that they use for balance. On their hands are curved, retractable claws.

Habitat:
The Sulidoror live in the northern mist country of the planet Belzagor, building primitive huts as a base for their hunting parties. Belzagor is sharply divided into tropical and arctic zones, with no temperate areas.

Culture:
The Sulidoror live a primitive existence, peacefully coexisting with the Nildoror, the dominant intelligent species on Belzagor. Deep in Sulidoror territory is the "Mountain of Rebirth," a religious shrine of the Nildoror, which is administered by the Sulidoror. The exact nature of the relationship between the Sulidoror and Nildoror is shrouded in mystery. They share the same language, and evidently have some means of rapid communication. Their philosophies are similar, and they seem to have the same religious beliefs.

The coat of an elder Sulidor. Sulidor fur deepens from rich mahogany to oxidized bronze with age.

Sulidor

THE THING_____

Source:
"Who Goes There?"
John W. Campbell
aka Don A. Stuart

Physical Characteristics:

The Thing is an entity made up of protoplasmic cells that can take any form at will. In its rarely-seen natural state, it is about 1.5 meters tall, with rubbery blue flesh and four armlike appendages ending in seven-tentacled hands. The pulpy head has three gleaming red eyes and thick writhing tendrils.

The Thing's intelligence does not reside in a single brain, but in every cell of its body. If even a few cells of the Thing are separated from it, they organize into an individual intelligent entity.

The Thing eats by absorbing the mass of its prey into its body, increasing in size by that amount.

Habitat:

The Thing prefers an environment where the temperature is about 50°C. It is not known what kind of atmosphere the Thing evolved in, since it is able to alter its body to breathe any gas. From the artificial light sources built by the Thing, it is assumed that it came from a planet orbiting a hot blue star.

During an exploratory expedition, Terran scientists discovered the remains of the Thing, frozen in polar ice.

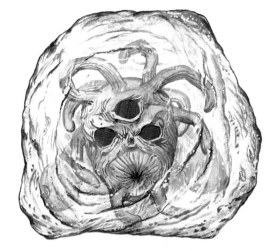

The Thing

THRINT _____

Source:
World of Ptavvs
Larry Niven

Physical Characteristics:

The Thrint are 1.25 meters tall, and are covered with iridescent green scales. At the corners of their broad, slashlike mouths they have thick gray tendrils, used for holding their food as they eat and manipulating small objects close to their eye. They are carnivorous, with needle-sharp metallic teeth.

Thrint are telepathic, and are able to impose their will on all other sentient species. This telepathic ability, known as the Power, gradually develops in the adolescent Thrint. If the Power does not manifest itself in a young Thrint, that Thrint is labeled a *Ptavv,* and is destroyed or sold into slavery.

Habitat:

The Thrint homeworld, Thrintun, is an earthlike planet, slightly smaller and of lower gravity than Terra.

Culture:

Many years ago, using their telepathic ability, the Thrint enslaved all other known sentient beings. Employing the technological skills of the slave races, they built a vast interstellar empire, often with a single Thrint family owning and ruling an entire slave planet.

Two to three billion years ago, one of the slave races, the Tunctipun, rebelled against the Thrint. Utilizing their skills in biological engineering, the Tunctipun developed semisentient species that were immune to Thrint control, and gave them to their masters. These animals turned on their Thrint owners and destroyed them.

All present knowledge of the Thrint comes from one Thrint who survived the war in a stasis field, and from artifacts and records put into stasis boxes during the collapse of the Thrint empire.

A sequential diagram of the fingerlike Thrint feeding tendrils in use (top view).

Thrint

TRAN

Source:
Icerigger
Alan Dean Foster

Physical Characteristics:

The Tran are two-meter-tall mammalian entities. Although they are very broad and well muscled, they are not strong, due to their light, hollow bones. Their long arms end in four-fingered hands.

The Tran's three-toed feet have long, curving claws that act as skates but that also can be retracted upward, with considerable effort, to allow it to walk on land. A small single claw on the Tran's heel acts as a brake.

The Tran move over the icy surface of their planet by wind-skating. Broad, membranous wings stretch from their hips to their arms, spreading as the arms are lifted to catch the wind.

With their slitted eyes, triangular ears, and short, thick fur, the Tran are somewhat feline in appearance. They are omnivorous beings, with both flat and pointed teeth.

Habitat:

The Tran inhabit Tran-ky-ky, an earthlike planet. The surface of Tran-ky-ky is covered with a shallow, frozen ocean dotted with archipelagos of small islands.

The average temperature at the equator is −15°C, but ranges to well below −100° toward the poles. The winds blow constantly, and can achieve velocities of up to 300 kilometers per hour.

Culture:

Most Tran live in small feudal island communities. They travel widely across their world on their claw skates, propelled by the wind in their wings. Other groups of Tran live a nomadlike existence, moving in a horde across the eternal ice and descending on the settled islanders to demand tribute of goods and food.

A Tran foot, showing skateblade claws in front and breaking claw in back.

Tran

TRIPED

Source:
Rule Golden
Damon Knight

Physical Characteristics:

The Triped is a trilaterally symmetrical entity with three legs, three arms, and six eyes functioning binocularly in three directions. The breathing orifice is located on the top of its head (covered in the illustration by an atmosphere-producing device). The Triped's three mouths are vertical slits, located between the three arms. Brain, heart, and lungs are all in the upper thorax, which expands evenly all around while breathing. The gracefully-curving spines around the neck stiffen with fear, trembling and relaxing as the entity becomes less afraid.

The Triped is telepathic, able to broadcast its emotions; it has no vocal cords.

Habitat:

The Tripeds live on a planet that orbits a star in the Terran constellation of Aquarius. Nothing more is known of their home world.

Culture:

The Tripeds are members of the Galactic Union, which administers peaceful relations between member cultures and species. Tripeds are vegetarians, and, along with other members of the Galactic Union, believe that carnivorousness and violence are traits that must be overcome before an intelligent species can attain true civilization.

The neck spines at rest (left) *indicate a calm Triped. The entity on the right is obviously agitated.*

This detail of a neck spine shows its fine hairs. It is hypothesized that these hairs somehow pick up telepathic signals.

Triped

TYREEAN

Source:
Up the Walls of the World
James Tiptree, Jr.

Physical Characteristics:

The Tyreeans are torpedo-shaped entities, with luminescent mantles that catch and glide on the constant high winds of the planet Tyree. Two muscular siphons provide propulsion.

The Tyreean can control the luminescent flashing of the mantle, creating subtly shaded colors to convey information. By merging their electromagnetic life force auras, complex and complete thoughts and memories can be passed from one Tyreean to another. The Tyreeans sense light and color as sound, and "see" the electromagnetic fields of other living things.

The small, active females are the dominant sex, exploring Tyree, gathering food, and creating safe home areas. The larger males have a more intense and stronger life field; they nurture and care for the children, sheltering infants in their pouches.

Habitat:

The Tyreeans live their lives in the high, strong upper gales of their planet. They build floating platforms, positioned on steady updrafts, for storage of possessions and food supplies. Tyree has a rich ecology of wind-borne plants and animals, which are herded and harvested by the Tyreeans. Tyreeans can sense the life force of other beings over interstellar distances, and have begun exploring other planets in this way from their permanent scientific bases at Tyree's two poles.

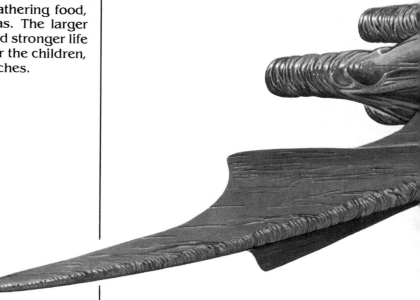

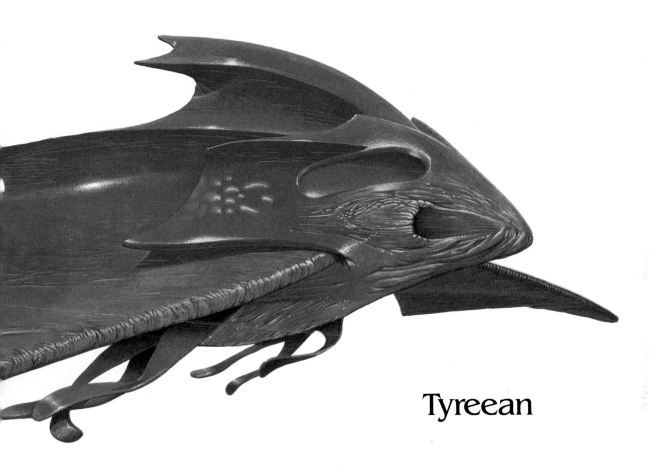

Tyreean

UCHJINIAN

Source:
Exiles at the Well of Souls
Jack L. Chalker

Physical Characteristics:

The Uchjinian is an intelligent, non-carbon-based entity shaped like an extremely pliable smear of matter, about one meter wide and two meters long. Individual Uchjinians range in color through the entire visible spectrum, but each of them is only one color. The Uchjinian is able to float freely in three dimensions, and move very quickly as its thick leading edge thins and contracts in any given direction.

The Uchjinians live underground during daylight hours, issuing from cracks in the ground at night. They are nearly invisible in twilight, becoming progressively brighter as darkness falls.

Habitat:

The Uchjinians live in an atmosphere that is mostly helium. They are completely non-technological, unable to make use of the simplest devices.

Nothing more is known about these elusive and unusual beings.

A side view of the mysterious Uchjinian.

Uchjinian

VEGAN

Source:
Have Spacesuit Will Travel
Robert A. Heinlein

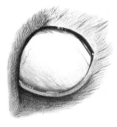

The Vegan nictitating membrane.

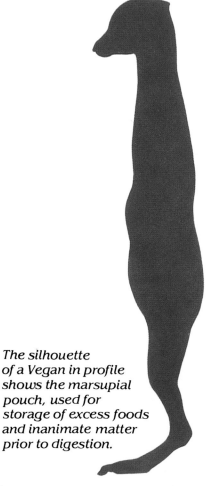

The silhouette of a Vegan in profile shows the marsupial pouch, used for storage of excess foods and inanimate matter prior to digestion.

Physical Characteristics:

The Vegan is about 1.2 meters tall, with a slender, extremely flexible body covered with creamy fur. Its large, prominent eyes have nictitating membranes. The hands have six fingers, all of which are mutually opposing. Vegans are quick and agile, and although they are not physically strong, their bodies are able to sustain massive damage without dying.

The Vegan language is musical, ranging from low to very high frequencies. They are able to convey the meaning of their words to beings who do not understand Vegan and to understand other languages by means of strong empathy bordering on telepathy.

Habitat:

Vegans developed on the fifth planet circling the star Vega. The planet's atmosphere is oxygen-based, but with a high level of ozone and nitrous oxide. Vega is a bright star, bathing its planets with high radiation and intense light.

Culture:

Vegans are members of the Three Galaxies, a loose organization of intelligent beings in the Milky Way and the Greater and Lesser Magellenic Clouds. Because of their high levels of inborn empathy for other races, the Vegans act as intergalactic police. They seek out and try to correct races that infringe on the rights of others.

Vegan

VELANTIAN

Source:
Children of the Lens
E. E. "Doc" Smith

Physical Characteristics:
The Velantian has a ten-meter-long serpentine body, strongly muscled and covered with almost impenetrable skin and scales. Talons on the six hands and feet are razor sharp, as is the knife-bladed tail. Broad, leathery wings used for gliding fold out of sight when the being is at rest.

Velantians are telepathic, and can project their thoughts even to nontelepathic races over enormous distances. Their brilliant and complex minds, their ability to withstand high gravitational stresses and accelerations, and their advanced scientific knowledge make them well suited to interstellar exploration and contact.

Habitat:
The Velantians inhabit the third planet of the star Velantia. Their planet, also called Velantia, is earthlike in all major respects.

Culture:
Early in their development, the Velantians were enslaved by the Delgonians, the inhabitants of the second planet in their system. The Delgonians, also telepathic, fed on the life force of tortured Velantians. In an age-long struggle to break free of the Delgon Overlords, the Velantians developed their physical and mental sciences in secret, until they finally discovered an effective mind shield. When Delgon control was broken by the shield, the Velantians built interplanetary ships and fought a savage war of extermination with Delgon. In the end, they utterly destroyed the Delgonians.

Velantian

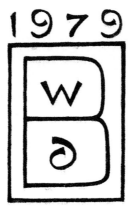

1979

THIS FOLIO OF PENCIL DRAWINGS IS TAKEN
DIRECTLY FROM MY SKETCHBOOK. HERE ARE
PREPARATORY RENDERINGS, NOTES, AND
STRUCTURAL, CUT-AWAY, AND LOCOMOTIVE
STUDIES OF EXTRATERRESTRIALS FOR THIS
BOOK, AND OF THYFE, A CHARACTER OF MY
OWN CREATION.

SALAMAN'S
HAND

SHOWING
EXCELLENT
OPPOSABLE
THUMB

SALAMAN
WHISTLING

SALAMEN MINUS UBIQUITOUS
PONCHO

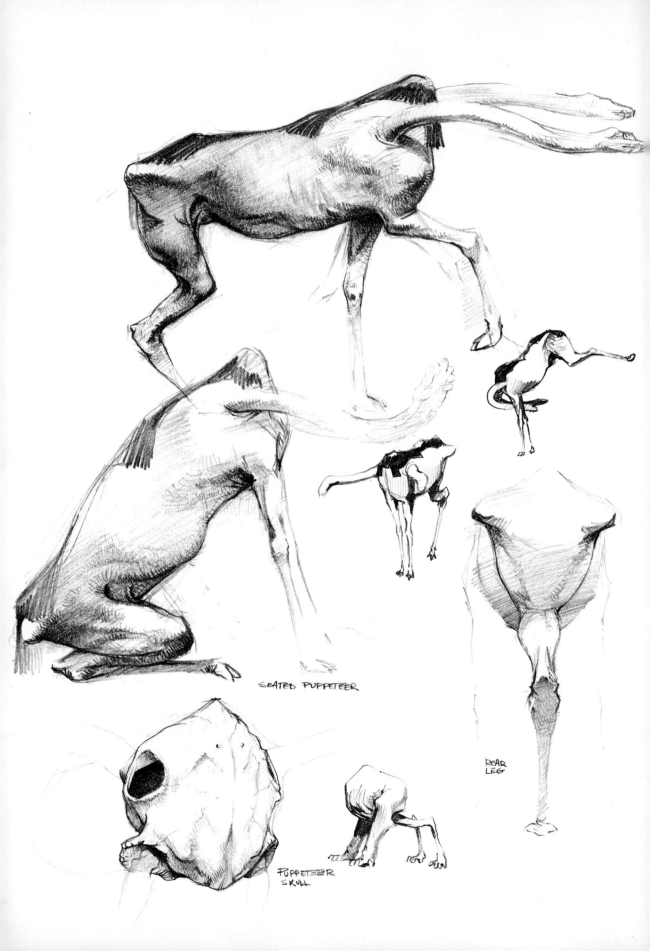

SEATED PUPPETEER

REAR
LEG

PUPPETEER
SKULL

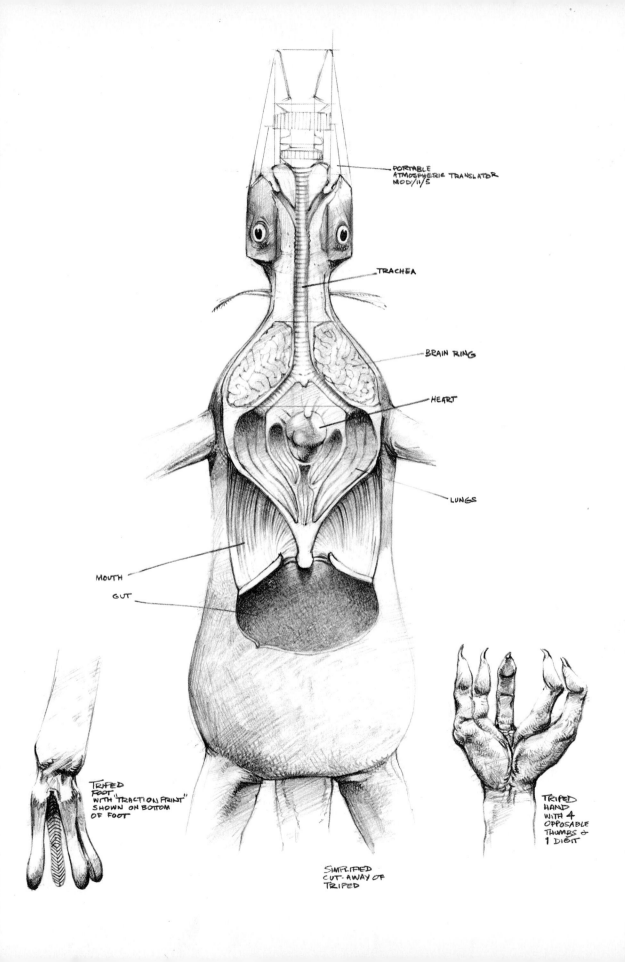

PORTABLE
ATMOSPHERIC TRANSLATOR
MOD/11/5

TRACHEA

BRAIN RING

HEART

LUNGS

MOUTH

GUT

TRIPED
FOOT
WITH "TRACTION PRINT"
SHOWN ON BOTTOM
OF FOOT

TRIPED
HAND
WITH 4
OPPOSABLE
THUMBS &
1 DIGIT

SIMPLIFIED
CUT-AWAY OF
TRIPED

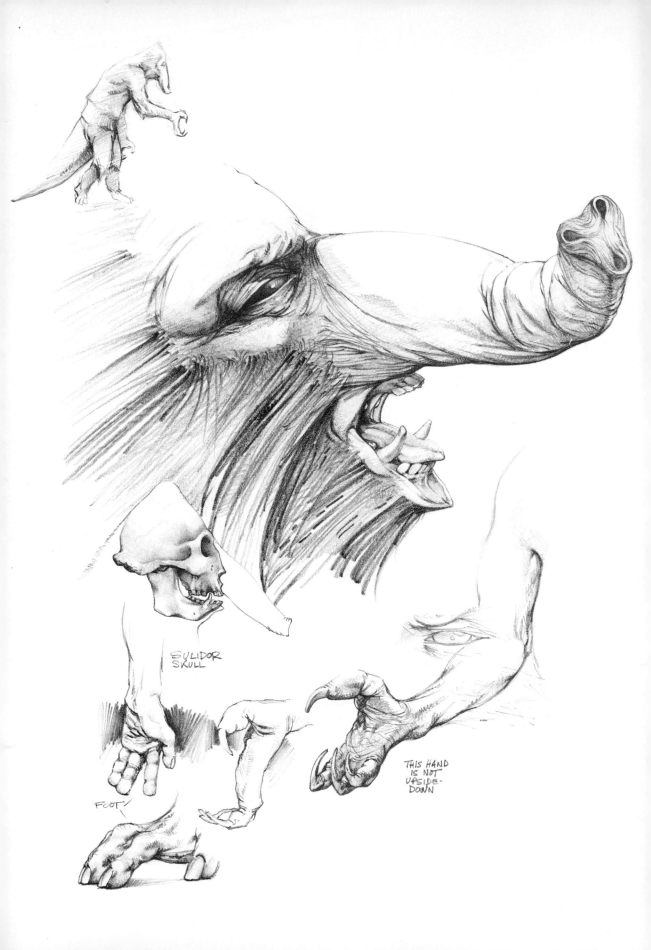

SULIDOR
SKULL

FOOT

THIS HAND
IS NOT
UPSIDE-
DOWN

TRAN

PUML DUELLING

MERSEIAN
SWORDSMAN

DEMON

THYPE IN ARMOR

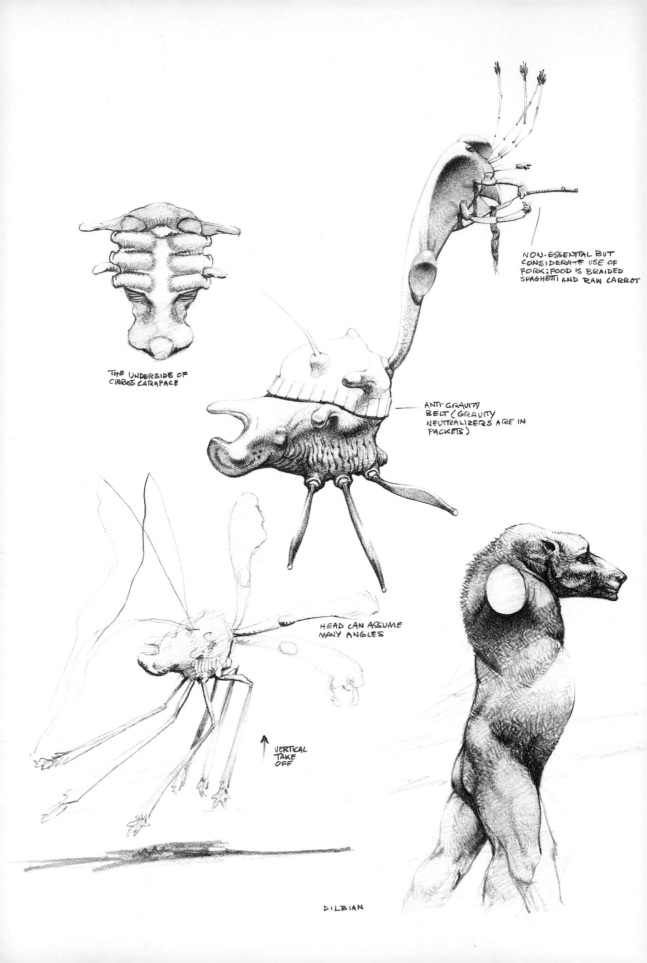

THE UNDERSIDE OF
CINRUSS CARAPACE

NON-ESSENTIAL BUT
CONSIDERATE USE OF
FORK; FOOD IS BRAIDED
SPAGHETTI AND RAW CARROT

ANTI-GRAVITY
BELT (GRAVITY
NEUTRALIZERS ARE IN
PACKETS)

HEAD CAN ASSUME
MANY ANGLES

VERTICAL
TAKE
OFF

DILBIAN

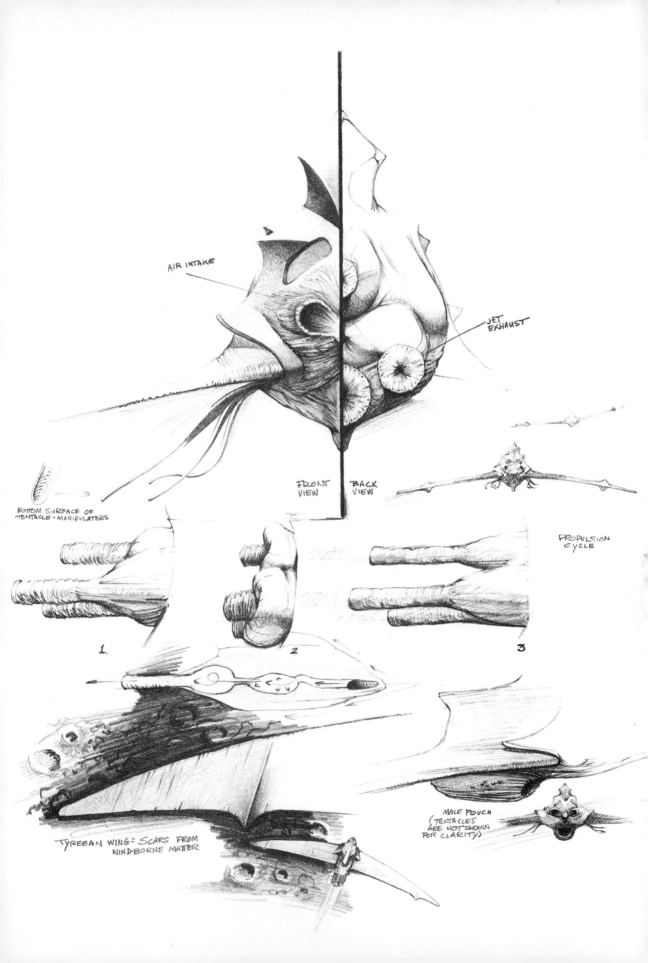

AIR INTAKE

JET EXHAUST

BOTTOM SURFACE OF
TENTACLE-MANIPULATORS

FRONT
VIEW

BACK
VIEW

PROPULSION
CYCLE

1.

2.

3.

TYREEAN WING: SCARS FROM
WINDBORNE MATTER

MALE POUCH
(TENTACLES
ARE NOT SHOWN
FOR CLARITY)

DEITY-FIGURE
FROM THYPE

THYPE ENCOUNTERING
A VILLAR

BEGGARS IN THE STREETS OF SITH

THYPE CROSSING THE PLAIN OF IB

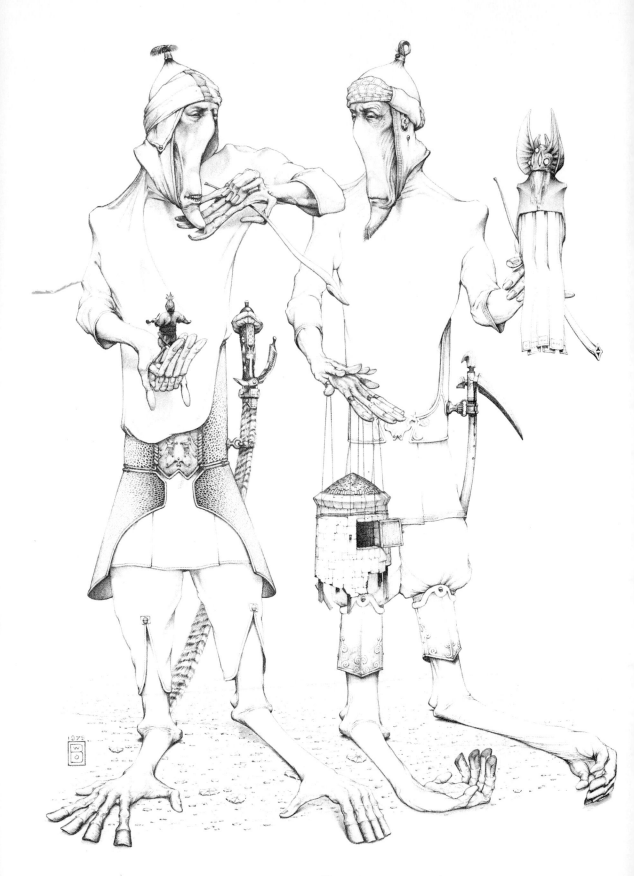

THYPE CAPTURED BY VASPARIAN NOMADS

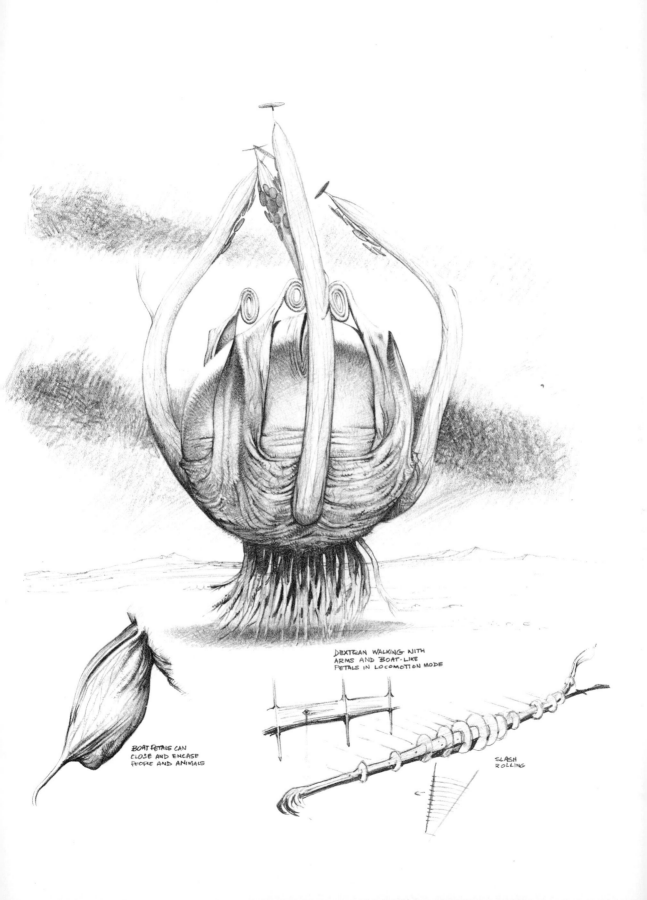

BOAT PETALS CAN
CLOSE AND ENCASE
PEOPLE AND ANIMALS

DEXTRAN WALKING WITH
ARMS AND BOAT-LIKE
PETALS IN LOCOMOTION MODE

SLASH
ROLLING

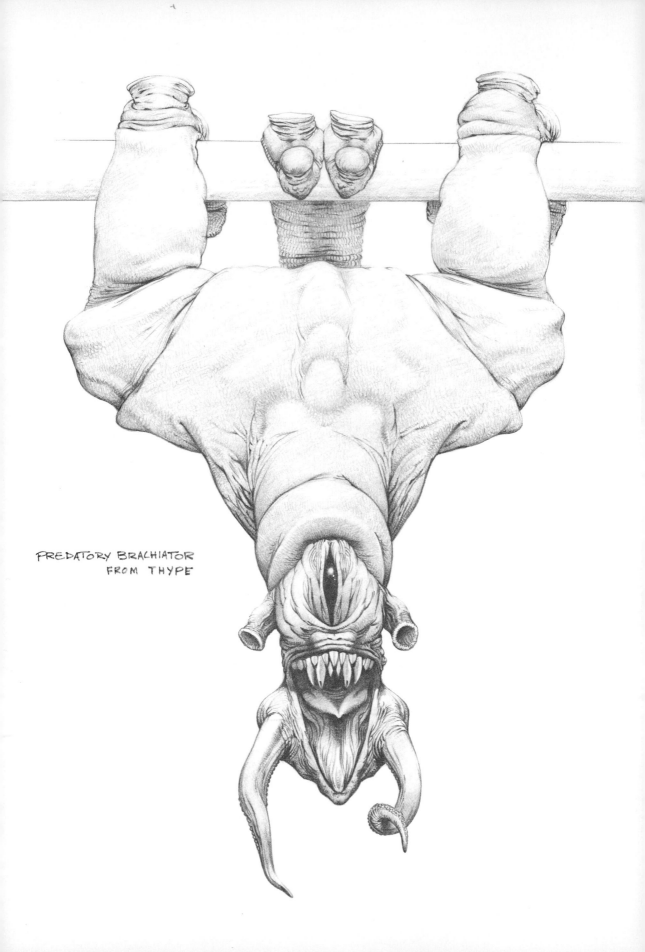

PREDATORY BRACHIATOR
FROM THYPE

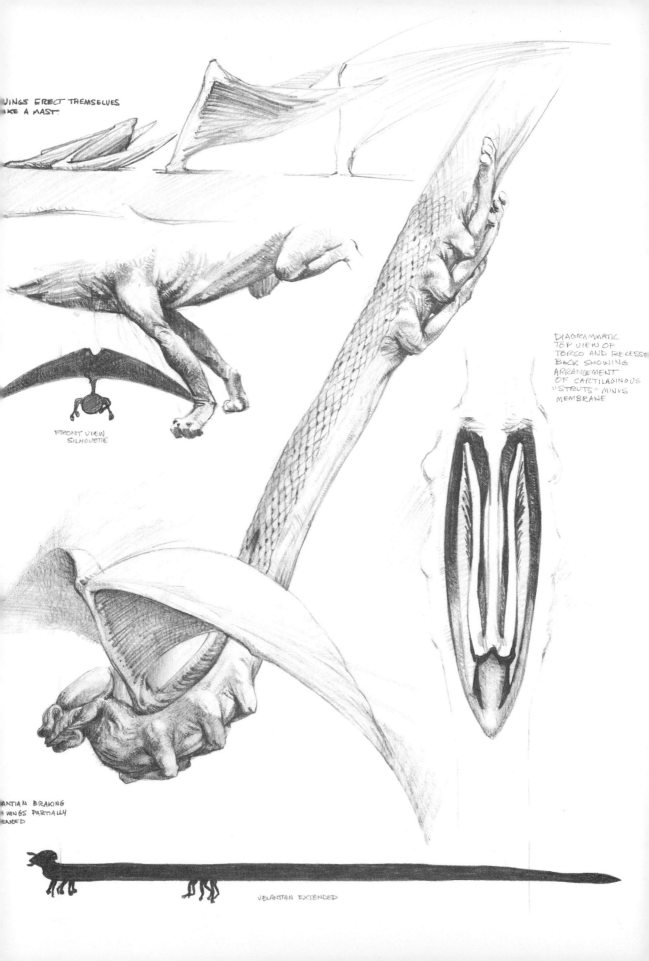

WINGS ERECT THEMSELVES
LIKE A MAST.

FRONT VIEW
SILHOUETTE

DIAGRAMMATIC
TOP VIEW OF
TORSO AND RECESSED
BACK SHOWING
ARRANGEMENT
OF CARTILAGINOUS
"STRUTS" MINUS
MEMBRANE

VELANTIAN BRAKING
WINGS PARTIALLY
EXTENDED

VELANTIAN EXTENDED

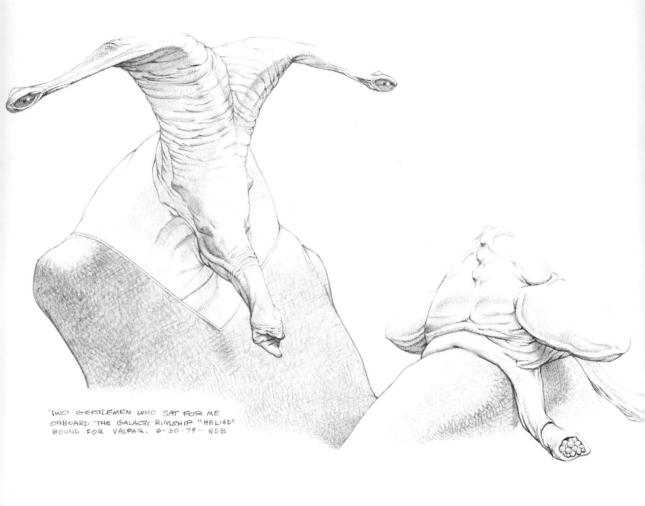

TWO GENTLEMEN WHO SAT FOR ME
ONBOARD THE GALACTIC RIMSHIP "HELIAD"
BOUND FOR VASPAR. 6·30·79 — WDB

TOP IS REALLY
FRONT OR
CHEST

HEAD
SWIVELS
AROUND

1   2

3

WITHOUT
CLOAK, ETC.
(DRAWN FROM
DEAD SPECIMAN)

PNUME AT REST & STANDING

DIRDIR

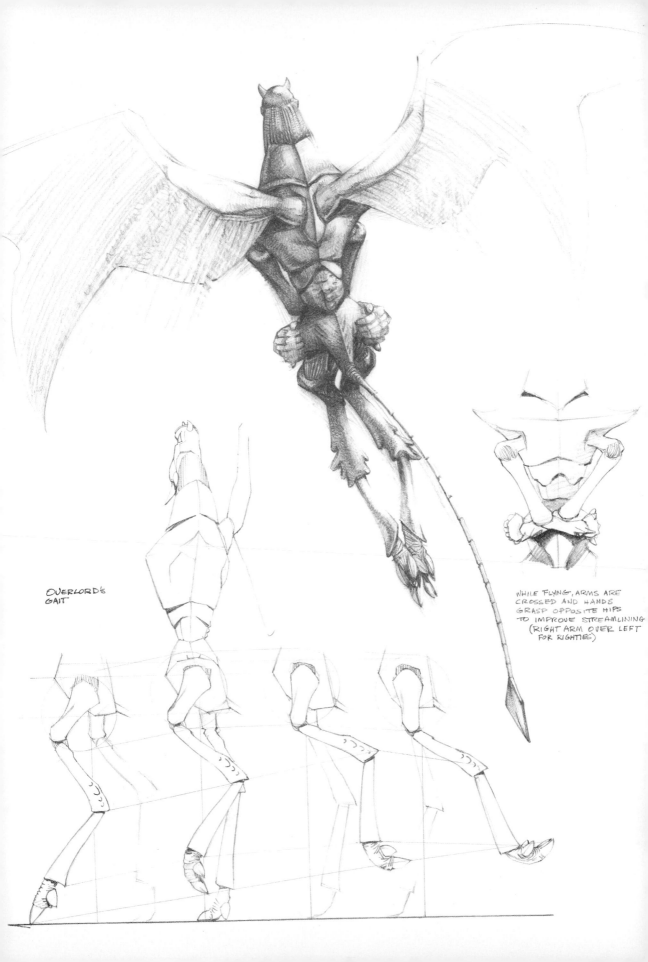

OVERLORD'S
GAIT

WHILE FLYING, ARMS ARE
CROSSED AND HANDS
GRASP OPPOSITE HIPS
TO IMPROVE STREAMLINING
(RIGHT ARM OVER LEFT
FOR RIGHTIES)

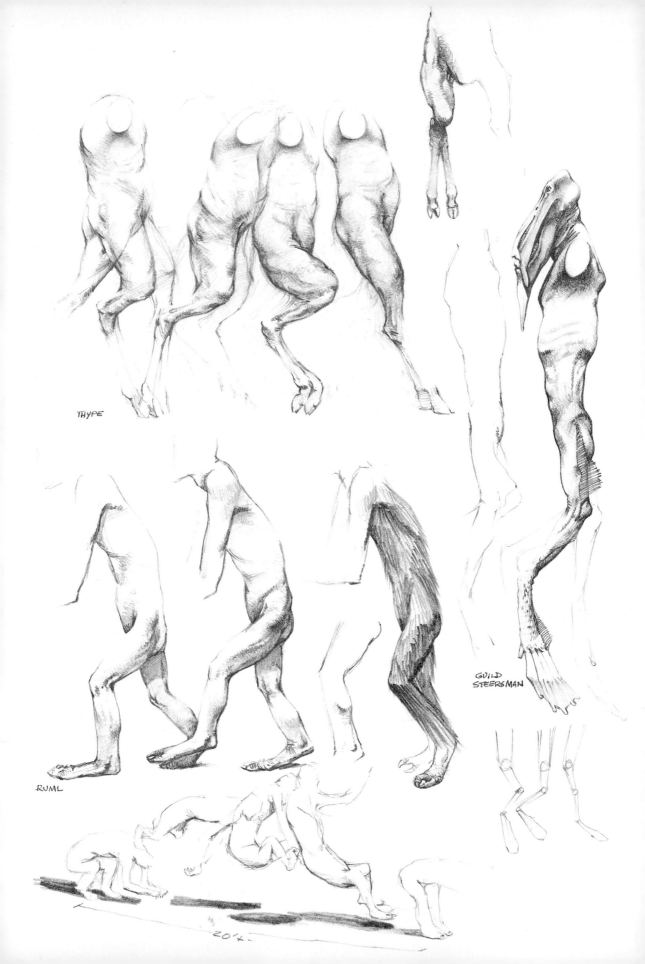

THYPE

RUML

GUILD
STEERSMAN

20'4

NECK
STRETCHES

BODY ATTENUATES
SOMEWHAT WHILE
LEAPING

IXTL JUMPING

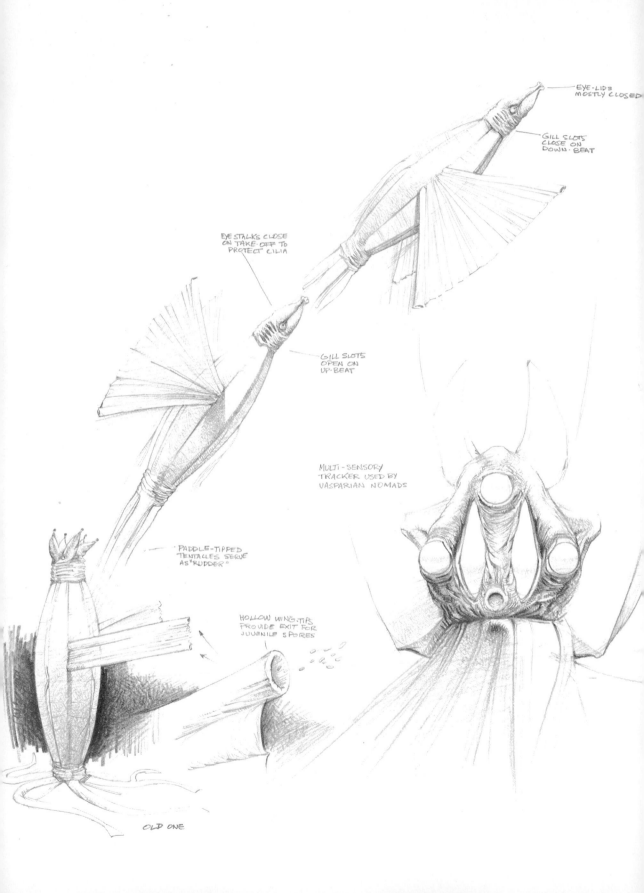

EYE-LIDS
MOSTLY CLOSED

GILL SLOTS
CLOSE ON
DOWN-BEAT

EYE STALKS CLOSE
ON TAKE-OFF TO
PROTECT CILIA

GILL SLOTS
OPEN ON
UP-BEAT

MULTI-SENSORY
TRACKER USED BY
VASPARIAN NOMADS

PADDLE-TIPPED
TENTACLES SERVE
AS "RUDDER"

HOLLOW WING-TIPS
PROVIDE EXIT FOR
JUVENILE SPORES

OLD ONE

GOWACHIN
SWIMMING IN BREEDING
POOL

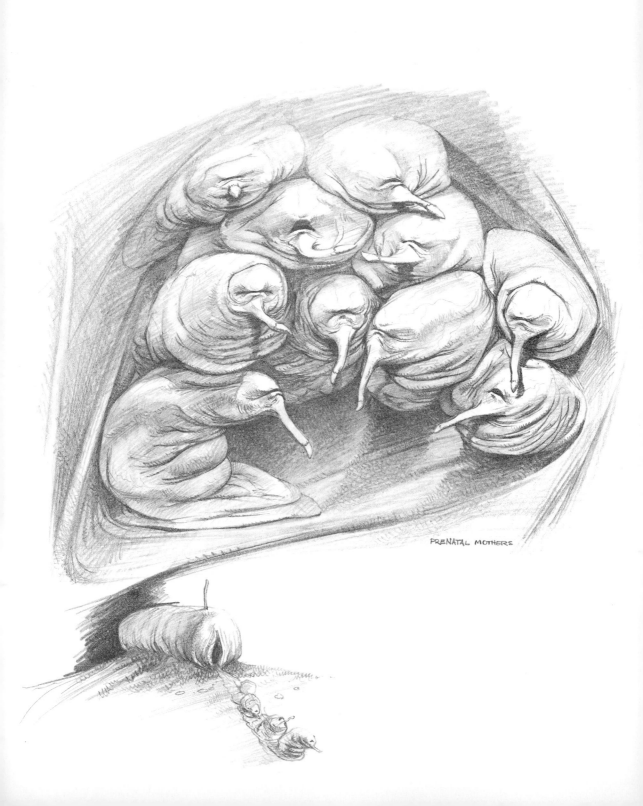

PRENATAL MOTHERS

REGUL'S POWER SLED, WITH
CONTROLS LOCATED IN CENTER
AND MANIPULATED BY TOES
(POWER-SLED INCORPORATES
HOVERCRAFT AND TRACTOR)

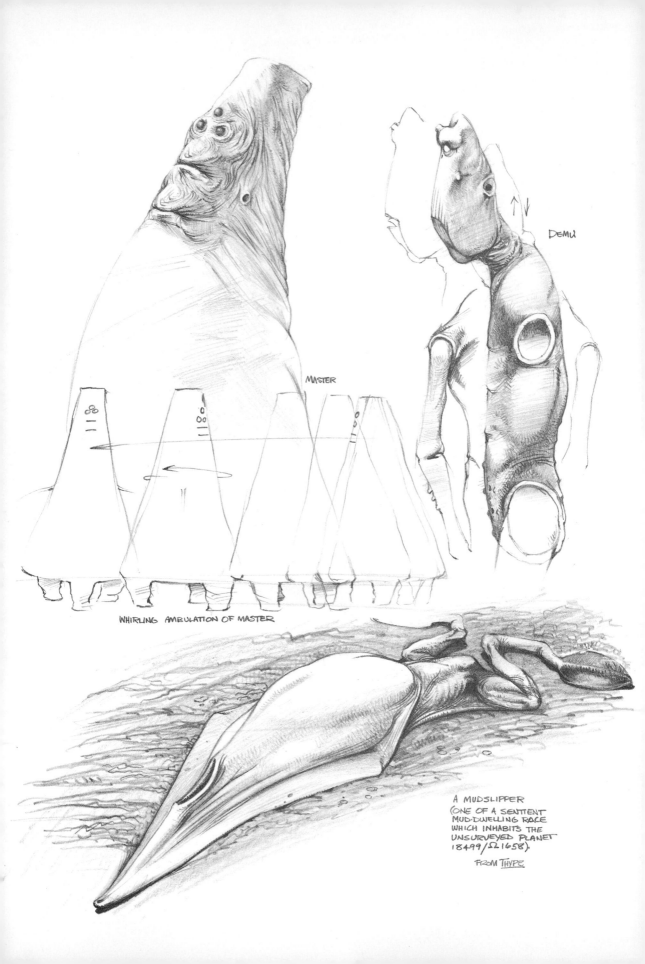

MASTER

WHIRLING AMBULATION OF MASTER

DEMU

A MUDSLIPPER
(ONE OF A SENTIENT
MUD-DWELLING RACE
WHICH INHABITS THE
UNSURVEYED PLANET
18499/Ω1658).

FROM THYPE

RIIM PARTHENOGENETIC PROCESS
IN LATE STAGE

Wayne Douglas Barlowe is the John James Audubon of otherworld creatures. He attended the Art Students League and Cooper Union, and apprenticed in the exhibition department at the American Museum of Natural History in New York. His illustrations have been seen on myriads of paperback books, and appear in *Tomorrow and Beyond*. Barlowe is currently at work on his own illustrated novel, called *Thype*. *Barlowe's Guide to Extraterrestrials* is his first book. His studio is located in Massapequa, Long Island.

Ian Summers, considered one of the most innovative art directors in the publishing industry, has art directed and designed over 2,000 paperback covers and scores of picture books. Summers has won the coveted Gold Medal from the Society of Illustrators and countless awards from *Art Direction* magazine, AIGA, and Neographics. His articles have appeared in *Madison Avenue*, *Art Direction* magazine, and *Publishers Weekly*. He has lectured at many science fiction conventions and teaches a special course in science fiction and fantasy illustration at the New School for Social Research.

Ian Summers is the author of *The Yearbook Book* and *The Fantastic Art of the Brothers Hildebrandt*, and editor of *Tomorrow and Beyond*. He produced the Brothers Hildebrandt's new novel, *Urshurak*. Summers resides with his wife and two children in Teaneck, New Jersey.

Wayne Douglas Barlowe

Ian Summers